Praise for *Vegetable Garden Tools*

Finally, a concise guide that gets right into the meat of it with zero fluff. Whether you're a veteran gardener that hasn't yet adopted the tools and tricks of the modern market gardener or a small farmer looking to intensify production this handy guide is sure to quickly prove its worth dozens of times over in the first season alone.

— Ben Falk, author, *The Resilient Farm and Homestead*

This is the first book on this topic that I have seen for decades. It is a book which all growers will dip into whether they are commercial growers or home gardeners. I appreciate the excellent illustrations and clear instructions.

— Rowe Morrow, Blue Mountains Permaculture Institute,
author, *Earth Restorer's Guide to Permaculture*

A nice little guide to many of the basic and specialized tools used by market growers today.

—Josh Volk, Slow Hand Farm, author, *Compact Farms* and *Build Your Own Farm Tools*

Praise for the Grower's Guides from the Market Gardener series

Jean-Martin Fortier has done it again! Incredibly beautiful, this series of pragmatic and professional instructional manuals makes it easy for anyone to start market gardening at a professional and regenerative level with a tailored focus on each aspect.

—Matt Powers, author, *Regenerative Soil* and *Regenerative Soil Microscopy*

Applying a market gardening mindset to your home garden will improve your yields and greatly reduce the amount of work involved. To do this, look no further than this series from Jean-Martin Fortier who is a pioneer in regenerative, biointensive, and organic market gardening.

—Rob and Michelle Avis, Verge Permaculture / 5th World,
co-authors, *Building Your Permaculture Property*

T0408184

GROWER'S GUIDES
Jean-Martin Fortier
FROM THE MARKET GARDENER

Vegetable Garden Tools

A Grower's Guide

TRANSLATED BY LAURIE BENNETT
EDITED BY PIERRE NESSMANN
ILLUSTRATIONS BY FLORE AVRAM

new society
PUBLISHERS

Copyright © 2025 by Jean-Martin Fortier. All rights reserved.
Translated by Laurie Bennett. Cover design by Diane McIntosh.
Printed in Canada. First printing May, 2025.

© Delachaux et Niestlé, Paris, 2023 First published in France under the title: *Les outils du potager. Les guides du jardinier-maraîcher* Jean-Martin Fortier, Flore Avram.

The author and publisher disclaim all responsibility for any liability, loss, or risk that may be associated with the application of any of the contents of this book.

Inquiries regarding requests to reprint all or part of *Vegetable Garden Tools* should be addressed to New Society Publishers at the address below. To order directly from the publishers, please call 250-247-9737 or order online at www.newsociety.com.

Any other inquiries can be directed by mail to:
New Society Publishers
P.O. Box 189, Gabriola Island, BC
V0R 1X0, Canada
(250) 247-9737

LIBRARY AND ARCHIVES CANADA CATALOGUING IN PUBLICATION
Title: Vegetable garden tools: a grower's guide / Jean-Martin Fortier; translated by Laurie Bennett; edited by Pierre Nessmann; illustrations by Flore Avram.
Other titles: Outils du potager. English
Names: Fortier, Jean-Martin, author | Bennett, Laurie, translator. | Nessmann, Pierre, editor | Avram, Flore, illustrator.
Description: Series statement: Grower's guides from the market gardener; 2 | Translation of: Les outils du potager.
Identifiers: Canadiana (print) 20250108097 | Canadiana (ebook) 20250108100 | ISBN 9781774060063 (softcover) | ISBN 9781550927993 (PDF) | ISBN 9781771423953 (EPUB)
Subjects: LCSH: Vegetable gardening equipment and supplies. | LCSH: Vegetable gardening | LCSH: truck farming.
Classification: LCC SB320.9 .F67 2025 | DDC 635.28—dc23

Funded by the Government of Canada | Financé par le gouvernement du Canada | Canada

New Society Publishers' mission is to publish books that contribute in fundamental ways to building an ecologically sustainable and just society, and to do so with the least possible impact on the environment, in a manner that models this vision.

EU COMPLIANCE PARTNER
new society PUBLISHERS
Certified B Corporation
MIX Paper | Supporting responsible forestry FSC www.fsc.org FSC® C016245

Presenting the collection

Grower's Guides from the Market Gardener

Hi!

I am delighted to bring you this new collection of practical guides. The advice you'll find in these books is based on working methods I developed on my own microfarm and refined over the last two decades. While plenty of these concepts are not new and were passed on to me by different mentors through the years, many other ideas stem from my own farming experience. I am sure you'll come across a number of tips and tricks that are innovative, proven, and easy to implement.

Whether you are a home gardener, hobby farmer, new market gardener, or an experienced farmer looking to transition to more intensive growing on smaller plots, you will find everything you need to take your horticultural practices even further.

Wishing you success and happiness in your agricultural adventures!

Jean-Martin Fortier, market gardener in Saint-Armand, Quebec

Creating a future where humans live in harmony with nature and with each other

Founded by Jean-Martin Fortier, the Market Gardener Institute is committed to inspiring and supporting new organic growers at every stage of their journey. Our mission is to equip them with the essential technical skills needed to thrive in their vital agricultural work.

Our vision is to multiply the number of organic, regenerative farms around the world and create a future where humans live in harmony with nature and each other.

www.themarketgardener.com

MARKET
GARDENER
INSTITUTE

Contents

Introduction: A Few Words About My Background 1
What Is the Market Gardener Method? 4
Preface: My Market Gardening Toolbox 8

PREPARING THE SOIL 11

Hand Tools 12
- Broadfork 13
- Bed Preparation Rake 18
- Wheel Hoe (Glaser) 24

Motorized Tools 26
- Tilther 27
- Walk-Behind Tractor 30

Tarps 34
- Silage Tarps 35

PLANTING AND SEEDING CROPS 41

Tools for Seeding Under Shelter 42
- Plug Flats 43
- The Paperpot System 47
- Dial Seed Sower 51

Tools for Seeding in the Field 54
- Jang Seeder 55
- Six-Row Seeder 58
- Seedbed Roller 60

Planting Equipment 62
- Row Markers 63
- Garden Trowel 67
- Paperpot Transplanter 68

Irrigation Equipment 70
- Drip Irrigation 73
- Sprinkler Irrigation 74

Crop Protection Tools 76
- Low Tunnels 77
- Caterpillar Tunnels 80
- High Tunnels 81
- Row Covers 83
- Insect Netting 86

CROP MANAGEMENT AND HARVEST 89

Weeding Tools 90
- Stirrup Hoe 91
- Collinear Hoe 94
- Wire Hoe 96
- Flex Tine Weeder 98
- Bio-Discs 102

Other Tools 104
- Hand Sprayer 105
- Backpack Sprayer 106
- Flame Weeder 108

Harvest Tools 110
- Quick Cut Greens Harvester 111
- Harvest Knife 114
- Pruning Shears 116
- Harvest Bins 117
- Harvest Bags 118
- Harvest Cart 119

Acknowledgments 120
About New Society Publishers 120

Introduction: A Few Words About My Background

Drawing on principles from agroecology, permaculture, and entrepreneurship, I champion a modern form of nonmechanized farming, carried out on a human scale.

On a human scale means feeding many local families, while respecting the human and natural ecosystems in which we operate.

On a human scale means allowing market gardeners to make a decent living from their work, to run their businesses as they see fit, and to give themselves more time off than conventional farmers.

On a human scale means evolving through the use of technology but especially by relying on people and their skills and knowledge.

From Organic Farms…

I studied agroecology at McGill University's School of Environment in Montréal, where I met my wife and business partner, Maude-Hélène Desroches. At the time, we were both looking to create a new model for farming, one that would have a positive environmental impact. After graduation, we spent two years in New Mexico, USA, working on an organic farm and learning to be market gardeners.

Our microfarming aspirations were later fueled by a trip to Cuba where we spent time on *organopónicos*, fascinating urban farms that were established during the American embargo. During that era, after the fall of the USSR, the country developed a biointensive and urban agricultural model to ensure food security for the island's residents.

...to a Family-Run Microfarm

Back in Quebec in 2004, we acquired a small plot of 10 acres in Saint-Armand, in the scenic Eastern Townships. On this land, we experimented with our innovative approach to market gardening, which especially drew from the work of Eliot Coleman, an American market gardener who has been highly influential in the world of organic microfarming.

We built a 2-acre market garden, Les Jardins de la Grelinette, where we were able to test the first iterations of my method, now called the Market Gardener Method. It consists of crop rotation, the near-exclusive use of hand tools, organic growing practices, and shorter marketing channels, with direct sales made through CSA boxes and farmers' markets. At Les Jardins de la Grelinette, Maude-Hélène and I both worked full-time, and hired two farm workers (one full-time and the other part-time) to help with harvests.

Making 2 Acres Profitable

Success came quickly, both in terms of harvests and direct sales. After bringing in $33,000 in our first year, we earned twice that in the following year, and more than $110,000 in our third year of operation.

We were thus able to earn a living as market gardeners from almost the very beginning. Since then, our farm has continued to feed more than 200 families every year, offering roughly 40 types of vegetables, all grown on just 2 acres. Over the years, our harvests expanded and sales continued to increase. Eight years after starting the farm, I presented this farming model in a practical guide called *The Market Gardener* in 2014. The book was an instant success—over 250,000 copies have now been sold, and it has been translated into nine languages.

In 2015, with the support of a generous patron, I founded Ferme des Quatre-Temps in Hemmingford, Québec, with the vision of creating a model for the future of ecological agriculture. On this 160-acre farm, we established a polyculture system in a closed-loop cycle, raising pasture-fed cattle, pigs, and hens, alongside a culinary laboratory. At the heart of the farm, 7.5 acres were dedicated to a market garden, where we applied the growing methods developed at Les Jardins de la Grelinette. It is here that I teach my apprentices the principles of productive and profitable market gardening.

The project was featured in a TV show called *Les fermiers*, which follows the evolution of Ferme des Quatre-Temps and its apprentices, who later start their own farms in front of the cameras. The show was a hit in Quebec and is now available on TV5 Monde and Apple TV.

In parallel, I worked to expand my methods to reach a broader, global audience. In 2018, we launched the Market Gardener Masterclass, a fully online course now available in over 90 countries. To further support this initiative, I founded the Market Gardener Institute with a clear mission: to educate the next generation of growers by equipping them with the knowledge, skills, and resources needed to become leaders in the organic farming movement.

The Institute has two key objectives: to teach best practices in market gardening techniques and growing methods, and to demonstrate that small-scale farming worldwide can not only be ecological but also productive and profitable. On a global scale, it's the number of farms, not their size, that holds the key to feeding the world.

Inspiring Change

My ambition is to drive meaningful change in society by promoting a way of farming that honors nature, supports communities, and empowers local farmers. I believe in a decentralized farming model, built farm by farm, as the foundation for a truly sustainable and resilient food system.

Since 2020, I have proudly served as an ambassador for the prestigious Rodale Institute, which researches regenerative organic farming practices in the United States and beyond. I am also honored to be the ambassador for Growers and Co., a company that develops tools and apparel for new organic growers.

What Is the Market Gardener Method?

While my approach may seem innovative, it is founded on practices that were first developed by nineteenth-century Parisian gardeners, who fed more than two million people through a network of thousands of market gardens—precursors to our modern-day microfarms—within the city of Paris.

These market gardeners applied remarkable ingenuity, skills, and knowledge to meet the increasing food demands of a city in the midst of urbanization and demographic expansion. They achieved this through organic, nonmechanized agriculture. From the mid-eighteenth century to the twentieth century, many books were written about the innovative practices of these market gardeners, whose technical feats were admired throughout Europe. But with the advent of modern practices, much of this know-how was relegated to the past.

As a result of mechanization, the advent of agronomic science, and improved refrigeration and transport that brought in fresh and inexpensive food grown abroad, farms grew in size, became less diversified, and took on a more technological focus—a trend that continues today.

Fortunately, these inspiring models led to the development of horticultural methods that have endured, and with the same objective: to grow sustainably, by maximizing vegetable yields without degrading soil quality. We now use the term "biointensive" to describe these methods. Unlike extensive agricultural operations, they continue to work on a human scale and offer farmers the opportunity to use little mechanization. Despite what some may believe, this approach is also profitable.

By working on only small plots of land, market gardeners can keep start-up investments to a minimum, compared to the funds needed for a conventional farm. Biointensive farmers also require a smaller workforce, doing the work themselves with the help of just a few employees. They also sell their produce directly to customers, avoiding commissions to intermediaries. These three factors allow market gardeners to start generating profits quickly.

Still, it's important to remember that working the land is never easy. While market gardeners can make a good living with this method, the first seasons are time-consuming and require a significant workload and financial investment. In this profession, nothing comes easy, and every dollar you earn is the fruit of your labor, the result of your organizational skills. That's why I always tell my apprentices to learn how to work smarter, not harder.

From a financial perspective, market gardeners should plan to start with an investment of $50,000 to $150,000, depending on whether certain assets are already available—such as a building that can be converted, access to abundant water, electricity, natural gas, or a vehicle. This amount does not include the cost of purchasing land, which can be amortized over 20 years, if needed. Renting is also an option that can prove very profitable, especially when the farm is located near a city or an affluent municipality, where land is expensive.

Regardless of experience and preparation, the first years of market gardening will be intense. Opening new ground, constructing greenhouses and tunnels, and setting up infrastructure (irrigation, washing and packing stations, nurseries, etc.) all take extra time and effort. However, once this phase is complete, market gardeners who have mastered their craft can do more than just make a living off a few acres—they can earn a very decent living.

This leads to another key principle I teach: your farm should work for you, not the other way around. Profitable and productive farming is possible, but you need to set it up for success.

Preface: My Market Gardening Toolbox

Using the "right" tools can make a huge difference in the success of a crop and, more broadly, in the success of a vegetable micro-farm. I put the word "right" in quotation marks here, because not all tools and equipment are equal when it comes to quality and function.

For more than two decades, I have been testing and using an array of small tools that have worked wonders in my gardens and in those of thousands of other market gardeners who have followed my lead. Many of the tools I use came from my mentors who, in my early days, encouraged me to discover the true value of using the right equipment.

In this little guide, I want to share some of my favorite tools with you, or at least the ones I think are a must-have for small-scale vegetable growers. They are used for soil preparation, seeding, planting, crop maintenance, harvests, etc.

These tool recommendations are specially aimed at small diversified vegetable farms, and, for the most part, they are perfectly suited to home gardeners and vegetable growers.

Whether you're a professional or a home gardener, the market gardening tools presented here will save you a lot of time, increase your productivity, and, above all, improve your ergonomics at work. In French, there is an old saying that goes, *les bons outils font le bon ouvrier*. It means "good tools make for a good worker," and I couldn't agree more! So, without further ado, here's a quick tour of my shed. I hope that some of these tools will become your best allies in your farming adventures.

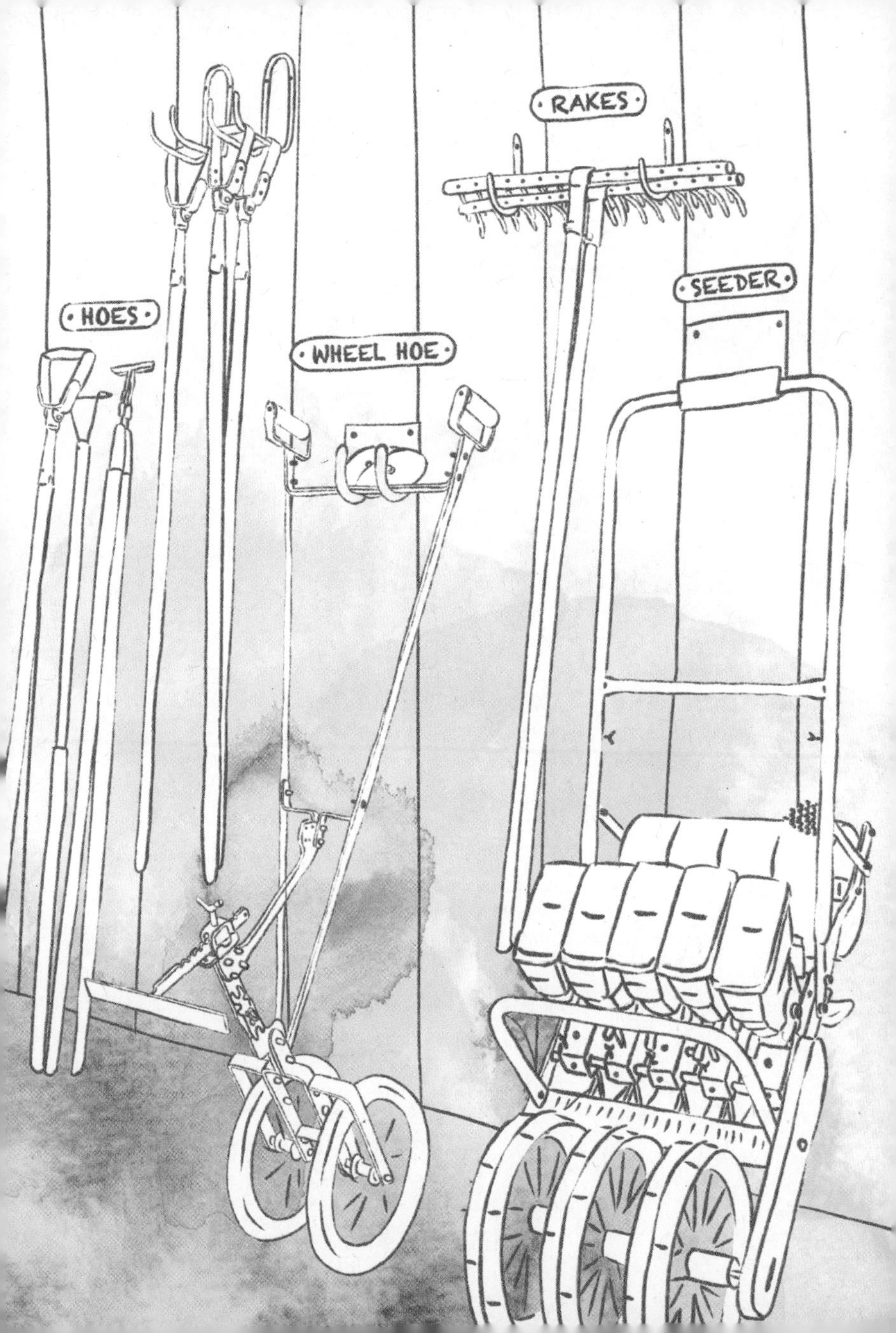
HOES
RAKES
WHEEL HOE
SEEDER

In biointensive market gardening, vegetable crops are seeded more densely and in a steady rhythm of crop successions. In practical terms, this means that beds are never left fallow; as soon as a crop reaches maturity, it is harvested and the soil is immediately prepared for the next seeding or planting. Working at this pace means that the soil is cultivated several times a year. To protect the balance of the soil and the microorganisms living within it, we avoid working the soil too deep, which would mix up various layers, as is the case with rototillers. Although this equipment is in the market gardener's toolbox, it is used sparingly. The soil is therefore worked only at the surface, using manual or battery-powered tools that loosen it to a depth of roughly 2 inches (5 cm). Once or twice a year, beds can be loosened at a greater depth—but this must be done without turning over the soil. To avoid disturbing soil life, you can employ the famous broadfork!

Preparing the Soil

Hand Tools

The permanent beds recommended in the Market Gardener Method are characterized by very precise dimensions, which are especially suited to the use of hand tools, operated by market gardeners leveraging their strength. Unlike tractor-mounted equipment that requires a bed width of at least 47 inches (120 cm), the hand tools recommended by Jean-Martin Fortier are perfectly suited to beds no wider than 30 inches (75 cm). With these tools, growers can loosen the soil before planting or seeding and then carry out various crop maintenance operations, such as weeding. All these operations take a gentle approach without compacting the soil, with the utmost respect for the microorganisms thriving underground, and with one of the fundamental principles of the Market Gardener Method—minimum tillage.

Broadfork

The broadfork, an ergonomic, easy-to-handle tool, is emblematic of biointensive market gardening. That's why Jean-Martin Fortier named his first farm, Les Jardins de la Grelinette, after the French word for broadfork, *grelinette*.

The History of the Broadfork

The original broadfork was invented by André Grelin (1906–1982), a French horticulturist and nurseryman who founded the Graines Grelin Frères seed company, with his son Olivier. In 1956, the tool won the first prize at the Concours Lépine, an innovation contest. It was later patented, in 1963, and then returned to the public domain in the 1980s. Today, the family's great-grandchildren continue to sell the original Grelin model, direct, which is still made in Savoie.

This authentic *grelinette* features hardened steel tines, a quality guarantee, and handles made from ash, which is a relatively flexible and highly durable wood. Other broadfork models have come on the market since the patent expired.

Purpose and Advantages

The broadfork gently loosens the surface soil without compacting it and, above all, without mixing soil layers, unlike rototillers and spades, which turn them over. As a result, the broadfork preserves insect life—especially worms, which are so important in keeping the soil alive—as well as bacteria and fungi, the very lifeblood of your soil.

As it works the soil, the tool also aerates it, providing the oxygen that these organisms require.

With a U-shaped handle and steel tines, the broadfork is an essential tool for soil preparation and maintenance: there is no better tool for this task. Ergonomic design helps to prevent user injury. Growers are less likely to experience back strain, do not need to use excessive force, and have a lower risk of developing blisters on their hands. Unlike spades, broadforks are leveraged with only the user's arms and body weight.

Easy to handle, broadforks are also quick and efficient. They drastically reduce time spent working the soil, which is typically quite hard on the joints when using more common tools. Another significant feature is that the broadfork can be used to pull up weeds and even harvest root vegetables like carrots and parsnips, as well as potatoes.

How to Use it

1 If the soil contains too many rocks and pebbles, it's best to first remove them with a rake to make the job easier.

Once the soil is ready to be loosened, keep your back straight while using your feet to drive the broadfork tines down into the soil.

2 Pivot the handles towards you, using them as a lever, to create a 45-degree angle between the tool and the ground.

3 Pull the broadfork out of the soil, while gently wiggling it from left to right. Step back about 8 inches (20 cm) and repeat.

The *campagnole*, a variation on the *grelinette* broadfork.

Maintaining Your Broadfork

To keep your broadfork in good shape, all you need to do is clean the tool and store it under shelter after each use. With these simple precautions, you can prevent rust and keep the wooden handle from deteriorating, especially in wet weather.

Different Models and Their Applications

Several broadfork models are available. The main variations on the market feature the number and size of tines, degree of sturdiness, and length of handles. The best model will depend on the type of soil and the size of your garden or farm.

→ FOR LIGHT SOILS (LOAM, SAND, ETC.)
Opt for a lightweight tool with a wooden handle and a tine count suited to the size of your operation: 3 tines for gardens up to 120 square yards (100 m^2), 4 tines for medium-sized operations, and 5 tines for 2.5 acres (1 ha) and up. At the end of winter, to loosen the soil in depth, nothing is more effective than running a broadfork down the bed, then amending the soil with compost, composted manure, or castor meal.

→ FOR HEAVY SOILS (ROCKY, CLAY CONTENT, ETC.)
Opt for a sturdier steel handle and a model with 3 or 4 fairly long tines. Note that a 3-tine model is more versatile and can also be used to get in between plants.

Over the years, variants on the original patented *grelinette* were developed under different names. For example, in the rotary broadfork category, the gardener turns the soil by rotating from left to right. The *campagnole* (pictured left), developed by Fabriculture in France, features two wheels that make it easier to move, and a mechanism that breaks up clods of dirt that rise to the surface. It is, however, more expensive than the classic broadfork.

Tip from Jean-Martin Fortier: Why and When You Should Work the Soil

Your soil should always be loosened beyond the surface layer because crop roots must be able to grow to a depth of 8 inches (20 cm). The soil also needs air to provide the right conditions for biological activity. Without these prerequisites, roots will not grow and nitrogen mineralization, which occurs when organic matter is broken down by bacteria and fungi, will not occur. To determine whether you should use a broadfork on a bed, you have to assess the state of the soil.

When the soil is in good condition, it can be cultivated at just the surface level, without the use of a broadfork. But when the soil is too compacted, it will require deeper work before planting, which is where the broadfork is effective.

Overall, the permanent bed system is less prone to soil compaction.

Lastly, to ensure the success of your seedlings, the soil surface should be cultivated with a tool like a power harrow (see p. 32).

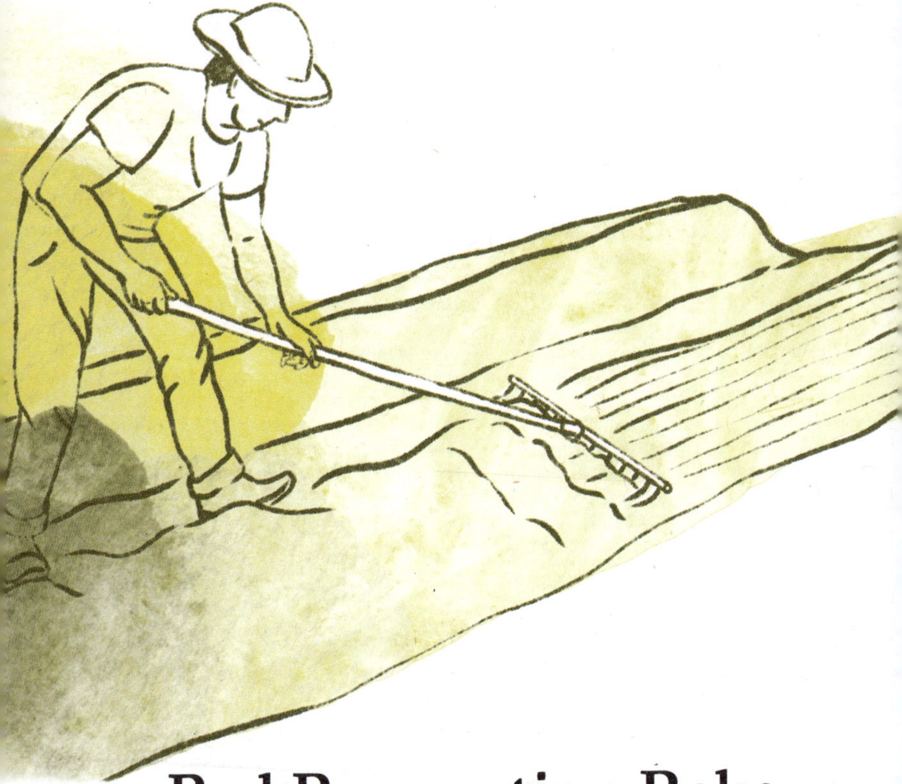

Bed Preparation Rake

The bed preparation rake, another essential tool for vegetable microfarmers, is primarily used to prepare beds. But it can also be fitted with accessories to serve as a clever row marker.

Purpose and Advantages

Bed preparation rakes, not to be confused with simple garden rakes, have been specially designed to prepare beds and to create uniform seeding surfaces on small-scale market gardens. They are flat and feature roughly 20 rigid tines, each about 4 inches (10 cm) long. Made of wood and aluminum, they are lightweight and easy to use. It is best to choose a bed preparation rake that is the same width as your beds, i.e., 30 inches (75 cm), if you are following Jean-Martin Fortier's recommendations.

The rake's purpose is threefold:

1. From the surface, it removes plant debris and rocks that could get caught in the seeder.
2. It levels the compost and amendments applied to permanent beds, thus preserving soil health and helping to maximize overall yields.
3. When plastic tubes are slipped onto the tines, the rake can mark rows for seeding or planting.

Tip from Jean-Martin Fortier

Use the back of the rake to spread and level organic soil amendments. To prevent injury, make sure to always keep your back straight.

How to Use It

1. Starting in the aisle, push the rake just a few inches into the soil, to avoid pulling too many rocks to the surface.

2. The tines should be right in front of you, with the top of the handle at chest height.

3. While keeping your back straight, walk backwards as you draw the rake diagonally across the bed, to bring all the debris towards the aisle. Walk only in the aisles.

Tip from Jean-Martin Fortier: Marking Rows for Seeding or Planting

By slipping flexible plastic tubes, known as row markers, over the tines, you can easily and cleverly turn your rake into a row marker. This lesser-known use of the rake eliminates the need for more expensive row marker rollers. To do this:

1. Slip plastic tubes, each roughly 8 inches (20 cm) long, onto the tines that line up with your intended row spacing.
2. Drag the rake down the length of the bed to create furrows. Be sure to form straight lines, as this will make weeding easier later on.
3. To create a grid, repeat the operation while dragging the rake across the width of the bed. Where the lines intersect is where you will know to seed or transplant.

Four Good Reasons to Work with Permanent Beds

GOOD DRAINAGE
Because the beds are raised, water easily drains from the surface. This means the soil can warm up faster in the spring and is less affected by leeching after heavy rainfall.

LESS SOIL COMPACTION
The aisles along the beds are specifically designed to allow market gardeners, and their wheelbarrows, to get around the farm. By avoiding trampling over growing surfaces, you can limit soil compaction. The soil is also better aerated and drains more effectively. By adding ramial woodchips made from small tree branches, you can turn the bed into an ideal host for microorganisms that will later benefit your crops.

OPTIMIZED GROWING SURFACES
Unsurprisingly, permanent beds are...permanent!

There are several advantages to always working the same soil area. First, it results in considerable improvements in soil quality. With this approach, you also know exactly how much material and time is needed to maintain the beds. This means you can optimize expenses and labor hours.

GENTLE WORK
Working on permanent beds requires no plows, no large harrows, and no subsoiling equipment, and therefore no tractors. This significantly reduces farm expenses, while preventing soil compaction. The manually operated tools presented in this book are designed for a permanent bed system.

Wheel Hoe (Glaser)

Wheel hoes, also known as oscillating wheel hoes and simply Glasers, serve a dual purpose: they loosen the soil before planting and act as a formidable weeding tool.

Purpose and Advantages

The wheel hoe consists of one wheel, two handles, and a sharp blade. It can replace the reciprocating spader and works without disturbing soil layers. The wheel hoe provides a gentler, more superficial action than the spader. It can be used to weed (see p. 102) between densely spaced plants in beds and to weed aisles. It can even be used to quickly and efficiently remove a row of plants before a new crop goes into the bed. Lastly, the wheel hoe can be used to incorporate soil amendments.

The Market Gardener Method recommends a wheel hoe with a 12-inch (30 cm) blade. This tool can also be paired with attachments, which makes it even more useful. For instance, when the wheel hoe is fitted with a second wheel, it can straddle tightly spaced rows. If the blade is too long, it can be cut shorter, and it can be sharpened for greater efficiency.

How to Use It

→ TO LOOSEN THE SOIL SURFACE

1. **Adjust the handle height so that you can stand comfortably, with your back straight, then adjust the angle and depth of the blade to suit the task at hand.**

2. **Push and pull the hoe as you move along the bed. Since the blade is sharp at both the front and back, you can work in both directions.**

→ TO INCORPORATE SOIL AMENDMENTS
Run the hoe diagonally across the bed, moving it back and forth to mix the compost into the surface soil.

Motorized Tools

To reduce the workload for market gardeners, the Market Gardener Method does not prohibit the use of battery- and gas-powered tools. It's just a matter of using them reasonably and wisely. Battery-powered market gardening equipment like the tilther, designed for soil preparation, and the Quick-Cut Greens Harvester both rely on a rotating mechanism powered by a rechargeable battery. They require attaching an everyday drill to the gardening tool. The walk-behind tractor, on the other hand, serves multiple functions, as it can be equipped with a range of removable accessories designed to prepare beds, shred and incorporate green manures, and mow the aisles; it should never be used as a substitute for hand tools.

Tilther

The tilther, or powered cultivator, was specifically designed for microfarms by Eliot Coleman, one of the pioneers of organic farming.

Purpose and Advantages

Designed to work with beds that are less than 50 feet (16 m) long, the tilther is an ideal tool for growers working with no more than 1 acre (4,000 m²) of land. It is used to incorporate soil amendments and to shred residues from previous crops. Like the other tools, the tilther leaves soil life undisturbed and only works the top 2 inches (5 cm). It really is the preferred tool for creating uniform seedbeds.

The tilther features a stainless steel power harrow, an aluminum cover, and adjustable wooden handles. A battery-powered drill, attached to the cover, drives the tilther, and the user operates the tool via a cable connected to one of the handles. The harrow tines are only 2 inches (5 cm) long, to cause as little soil disturbance as possible. Because the entire head is 15 inches (38 cm) wide, the tilther is the perfect size to tackle beds that are up to 30 inches (75 cm) wide in one round trip.

Buying a tilther requires a small investment, but it quickly pays for itself in the time you save! For greater efficiency, we recommend a drill that provides at least 18 volts, which must be purchased separately.

Tip from Jean-Martin Fortier

The tilther is the ideal tool when you need an efficient alternative to the more expensive and heavy walk-behind tractor. What's more, since it is powered by a drill, it consumes no fossil fuels and makes very little noise. It is also convenient for working in small spaces like greenhouses and tunnels. It is lightweight, highly ergonomic, and much easier to handle than a walk-behind tractor. However, if your beds are more than 50 feet (15 m) long, a power harrow on a walk-behind tractor will be more effective, or a rototiller followed by a seedbed roller (see p. 60). In such cases, using the tilther, due to its small size, would be a waste of time.

How to Use It

1. Use a bed preparation rake to clear the bed of debris (see p. 18). This first step is essential because debris can clog up the tilther. If you've invested in the tool, you might as well treat it with care!

2. Loosen the soil with the broadfork. The tilther is more of a finishing tool, so it does not work well with harder soils. Then, amend the bed with an organic fertilizer.

3. Bring the tilther to one side of the bed, while standing in the aisle. Adjust the handles so that you can comfortably walk it down the entire length of the bed.

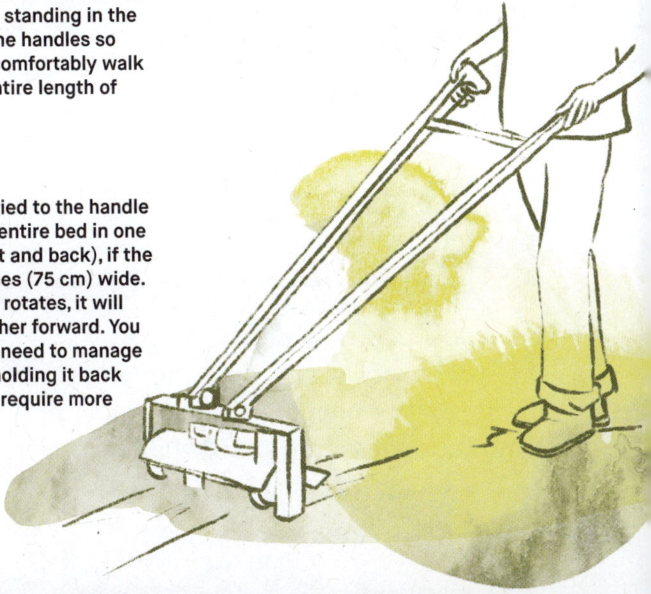

4. Pull the cord tied to the handle and work the entire bed in one round trip (out and back), if the bed is 30 inches (75 cm) wide. As the harrow rotates, it will propel the tilther forward. You will therefore need to manage its speed by holding it back if some areas require more attention.

Walk-Behind Tractor

The walk-behind tractor can be used with a wide range of accessories and is suitable for farms growing crops on more than 1 acre (4,000 m^2). It is more maneuverable and much less impactful than a conventional tractor.

Purpose and Advantages

For biointensive market gardening, the walk-behind tractor is the only useful piece of mechanized equipment. Although it is quite expensive, it quickly pays for itself with significant time savings. It requires little maintenance, and operating it is not particularly strenuous—an advantage that is worth noting, as this profession puts a lot of strain on the body!

The walk-behind tractor is a powerful machine that can be used in difficult terrain, on heavy soils, and on beds containing lots of plant residues and rocks, for instance.

Its greatest strength lies in its versatility. The walk-behind tractor can be fitted with a wide range of accessories. You can cultivate soil surfaces with a power harrow, mow green manures with a flail mower, and form beds with a bed shaper.

Another advantage of the walk-behind tractor is that it can be equipped with a roller at the back. This tool levels and lightly compacts the soil surface. Using the walk-behind tractor with these accessories ensures optimal contact between soil and seeds.

Walk-Behind Tractor or Rototiller?

Many growers choose to use a rototiller rather than a walk-behind tractor because it is less expensive. The problem is that it works the soil too deep and mixes soil layers. As a result, the rototiller generates a plow pan, also called hardpan, just below the tilling depth. Water tends to stagnate along the plow pan, and air passes through less easily. Roots will also have trouble growing through this layer, causing them to grow laterally instead, which is detrimental to plant development. If rototillers are used frequently or over extended periods of time, they result in compacted beds and soil that is not ideal for seeding and planting. Growers should therefore avoid using the tool more than once or twice per year in any given bed. After running the rototiller down a bed, use a seedbed roller to gently compact the soil, to prepare it for the next seeding or transplanting.

How to Use It

1. Attach the power harrow and roller to the walk-behind tractor.

2. Set the tine height to your preferred working depth, which should be no more than 2 inches (5 cm).

3. Stand in the aisle and adjust the handles so that you can comfortably operate the tool, keeping your back straight.

4. Start the walk-behind tractor and run it down the length of the bed. In a single pass, the soil will be tamped down and leveled. Any amendments will be thoroughly incorporated into the first few inches of soil.

Tip from Jean-Martin Fortier

The power harrow that attaches to the walk-behind tractor is specifically designed for 30-inch (75 cm) permanent beds used in biointensive market gardening. While loosening the soil, it also does a better job of leaving soil microorganisms undisturbed. With this tool, you can get a perfect bed surface, ready for the next direct seeding or transplant.

Jean-Martin Fortier's Preferred Model

My favorite walk-behind tractors are made by an Italian company called BCS. Select a model with long adjustable and reversible handles. With these features, operators can avoid stepping on the beds, instead rotating the handles to steer the tractor while walking in the aisle.

Tarps

I have experimented with several soil covering methods to prepare beds for the next crop. I initially attempted to mulch blocks with crop residues like compost, but this turned out to be quite tedious. Soon, I began using black polyethylene tarps. The results were decisive: covering the soil with this type of tarp led to a significant decrease in weed development in the next crop.

As soon as a vegetable crop has been harvested, the soil is immediately covered with a tarp that, by preventing any light from reaching the soil, speeds the breakdown of crop residue and neutralizes weeds. And this process does not harm or degrade soil life since the tarp keeps the ground moist and protects it from inclement weather, which in turn creates a welcoming environment for earthworms.

Silage Tarps

Silage tarps are inexpensive, quick to set up, highly effective, and if handled with care, they can be reused many times. They considerably reduce the need to work the soil.

Purpose and Advantages

Silage tarps made of UV treated polyethylene deliver spectacular results in biointensive market gardening. First, they speed up the composting of plant residues remaining on the beds. In just a few weeks, the surface will be ready for the next crop. Thus, silage tarps increase the yields of permanent beds by shortening the transition time between two successive crops. When sheltered from the sun and predators—like birds, for instance—earthworms, pill bugs, and other living organisms proliferate in moist soils. They quickly break down organic matter and crop residues in the ground.

The use of silage tarps, sometimes called "occultation," is also an effective way to manage weeds. The moist soil and warm surface temperatures are favorable for weeds, so they germinate easily, but quickly die off due to a lack of sunlight. Like the stale seedbed method, a silage tarp makes it possible to rid the bed of some weed seeds.

For smaller home vegetable gardens, straw often serves the same purpose as a tarp. But for microfarms, with beds covering half an acre or more, silage tarps are faster to set up and are more effective. Spreading straw over large areas is highly time consuming and can be ineffective. If the mulch thickness is not well managed, this layer will not fully block sunlight, resulting in beds that are not ready once the straw is removed.

These tarps are readily available and reasonably priced. The only drawback is that they are not easy to handle and are a little heavy. To avoid having to move them, it is best to store them alongside your field blocks when they are not in use. They have a lifespan of about ten years.

Occultation and Solarization: What's the Difference?

Occultation uses opaque tarps, and contrary to popular belief, silage tarps, although they are black and do absorb heat from the sun, do not warm the soil. The heat remains at the surface of the tarp. Solarization uses clear plastic; so if you need to warm up a bed, it is better to opt for a clear plastic tarp. Solarizing beds is quite risky in the summer, but you can do it, if necessary, at the start and end of the season.

Good to know: solarization can also destroy weeds, provided you keep the tarp on the bed for several weeks, in the heat of summer. Unfortunately, it is always hard to keep your beds inactive for that long in the middle of summer!

How to Use It

1. Once a permanent bed has been harvested and is ready for a new planting or seeding, you may want to use a walk-behind tractor fitted with a flail mower to mulch any green manure remaining on the soil surface.

2. Cover the bed with a silage tarp. If needed, wet the soil before covering it, so that microorganisms will find the soil hospitable.

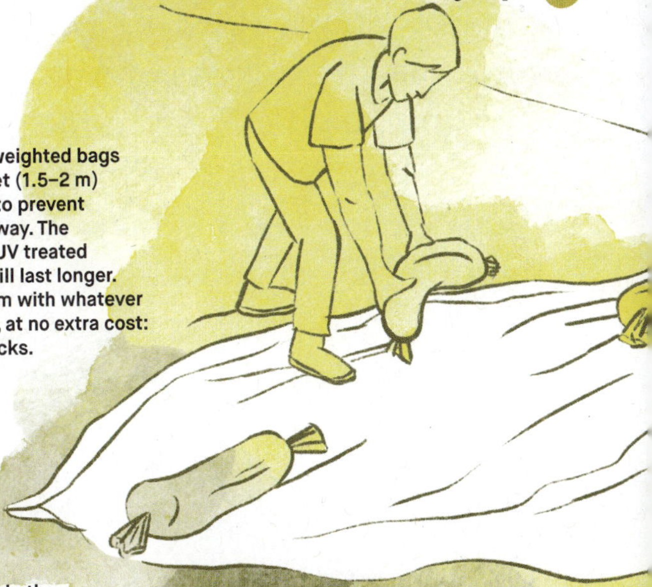

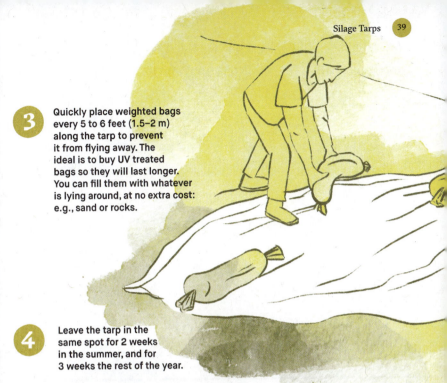

3 Quickly place weighted bags every 5 to 6 feet (1.5–2 m) along the tarp to prevent it from flying away. The ideal is to buy UV treated bags so they will last longer. You can fill them with whatever is lying around, at no extra cost: e.g., sand or rocks.

4 Leave the tarp in the same spot for 2 weeks in the summer, and for 3 weeks the rest of the year.

At the end of the season, covering the beds will provide two advantages: water will not accumulate in the soil over the winter, and the ground will be less prone to erosion. In the spring, the beds will be ready for the first seeding as soon as the tarps are removed! In French, this is sometimes called *un labour biologique*, which means "biological tilling" in English.

Tip from Jean-Martin Fortier

It really would be a shame not to use silage tarps because they save a lot of time and serve many valuable purposes:

- Clear up residue from previous crops remaining on permanent beds
- Prevent and even reduce weed growth
- Keep the soil moist, even when no crop is growing in the bed
- Protect the soil from leaching in heavy rain

While soil preparation is a crucial step that determines the success of a crop, starting plants and putting them in the ground is just as important, and requires special care. With intensive organic market gardening, extending the growing season, and therefore the harvest season—especially in the spring—involves starting and growing seedlings in a nursery and under cover. Later, they will be planted outside, in the field, and may need protection from inclement weather. Then, as soon as the weather is warm enough, crops will be sown and transplanted directly in the field without any cover. However it's done, seeding and transplanting crops involves the use of many tools that make it easier for market gardeners to get the job done: seeders, transplanters, and row markers. Not to mention setting up complementary equipment that is essential to crop success: irrigation systems and equipment that provides protection from inclement weather and pest damage.

Planting and Seeding Crops

Tools for Seeding Under Shelter

Growing your own vegetable seedlings in a nursery is both a great pleasure and the best way to control crop production from A to Z. It increases your odds of growing successful vegetables, makes your operation less dependent on outside farms, and allows you to better manage crop planning and successions. However, growing your own plants requires a fairly large investment in terms of both equipment and time, hence the need for a solid crop plan and, at times, allowing yourself to rely on other farms for your plant starts to help cope with a heavy workload.

Plug Flats

Plug flats make it much more efficient to grow seedlings under shelter. They use space effectively and make it easier to care for and monitor young plants.

Purpose and Advantages

Plug flats are made up of plastic cells that are connected along the edges. Before being transported, they are placed in a tray. They contain between 24 and 200 cells. The number of cells you need will depend on the root ball size you want to achieve and the amount of time seedlings will remain in the flat. These trays are easy to handle and fill with potting mix.

They save time especially when planting the seedlings. Since each root ball is in its own cell and cannot become entangled with others, it is easier to pull each plant out individually. Although the trays are quite strong and can be reused for several seasons, they will eventually fall apart and have to be replaced.

How to Use Them

1 Fill cells completely with potting mix and scrape off any excess soil with a stick or brush.

Next, lift the tray a few inches and then let it drop back down, to help the soil settle. Top up the cells, if needed.

2 Using your finger or a dibble board, make a depression in each cell to mark where to drop seeds.

3 By hand or with a seeder, drop a seed into each indentation made in step two. Then fill with potting mix and tamp lightly with the palm of your hand.

4 Water with a gentle spray, using a water breaker nozzle or a watering can fitted with a rose head.

Tip from Jean-Martin Fortier

A week before planting seedlings outdoors, it's best to harden them off by taking them out of the greenhouse. Of course, you will need to bring them inside in the event of a frost or very cold weather.

After planting the seedlings, it's important to clean the trays thoroughly. Once all traces of potting soil are gone, leave them out to dry in the sun for a few hours. This will prevent the proliferation of bacteria and disease in the cells, to ensure future seedlings are healthy.

Potting Mix

Potting mix is one of the fundamental elements in biointensive market gardening, so you need to choose it carefully. Seedlings require specific amounts of mineral elements that the potting mix must satisfy. It therefore has to meet certain criteria, such as porosity (for good drainage) and nutrient content. For beginner market gardeners, it is probably easier to buy commercial potting mix. For those who are looking to make their own potting mix, here is a simple and effective recipe for everyday use.

- 1.7 ft^3 (48 L) of quality peat (not too fine, not too coarse).
- 1.1 ft^3 (32 L) of perlite or vermiculite. It serves as an aggregate and regulates drainage and aeration.
- 1.1 ft^3 (32 L) of potting compost that is fertile and made from finished compost.
- 0.6 ft^3 (16 L) of good garden soil to counter the acidity of the compost and provide better structure. The soil you use must be light (not too sandy, not too much clay).
- 1 cup (240 ml) of blood meal, for nitrogen content. Can be replaced with feather meal.
- ½ cup (120 ml) of agricultural lime to adjust the pH, because peat is naturally acidic.

Pour the lime and peat into a wheelbarrow and mix them with a shovel before adding the other ingredients. To get the soil moist, lightly water the substrate while turning it over. Once you have a homogenous mix, sift it through a ½-inch wire mesh. If working with large quantities of potting mix, you can use a concrete mixer to save time.

The Paperpot System

When it comes to transplanting, the Paperpot system has been revolutionary in allowing growers to plant seedlings without having to remove each one from a pot or tray.

Purpose and Advantages

Paperpots, or paper chain pots, are strips of paper stuck together to create hexagonal cells. Each chain contains 264 cells. The manufacturer provides all the equipment needed to use the Paperpot system: rigid plastic seedling trays, a metal frame that supports the paper chains, spreader bars, the plastic seeder frame, and the transplanter itself (see p. 68). For microfarms, the complete starter kit is a small investment, but it's worth it. Paper chain pots are sold individually or in boxes of about 75 units.

Once the seedlings are ready to be transplanted, these honeycomb structures go right into the soil, using the manually operated Paperpot transplanter. They do not need to be individually pulled from their cells, because their biodegradable paper will gradually disintegrate in your soil. This is a huge time-saver.

Tip from Jean-Martin Fortier

The Paperpot Transplanter is the very best tool to grow and transplant seedlings efficiently. With a range of paper chain pot dimensions, you can adjust in-row spacing with precision. It makes the tedious planting stage twice as fast as it would be with plug flats. No more need to hire additional workers for transplanting.

One person is enough to get this step done. The Paperpot system is also ideal for growing early vegetables and crops typically sown right in the field, like turnips, radishes, and even beets.

How to Use It

1 Use the spreader bars provided by the manufacturer to stretch out the paper chain pots.

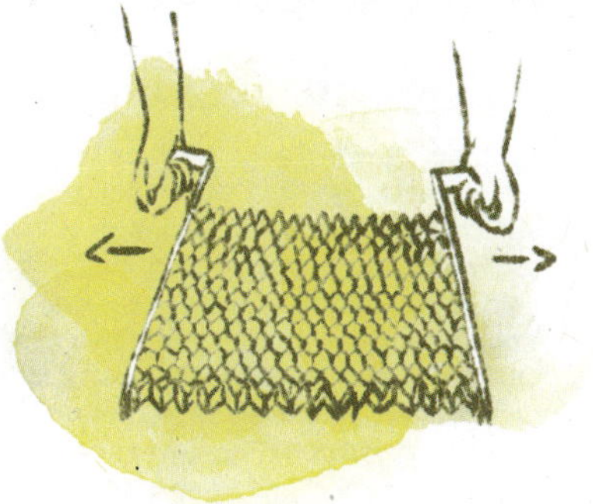

2 Fit the honeycomb grid onto the metal frame by sliding it onto the metal tabs at the edges. Make sure to line it up in the right direction.

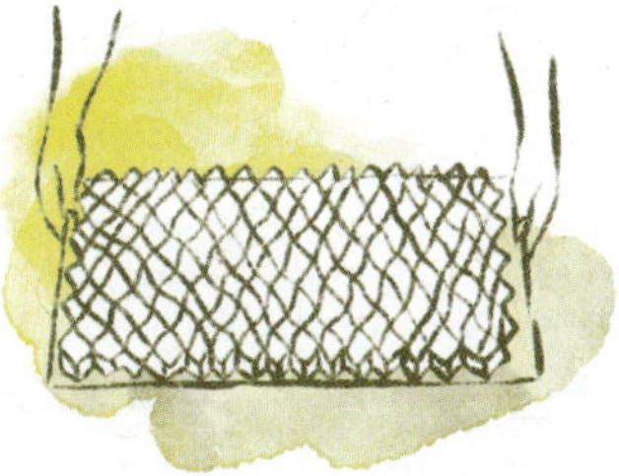

3 Fill the cells with potting mix, then seed as you would with a plug flat (see p. 43). Sprinkle with sifted potting mix to cover the seeds. Remove the metal frame, then water with a gentle spray. Clean the equipment thoroughly so that it can be used to prepare other seedlings.

Seedling Room or Nursery?

To start your growing season off on the right foot, seeding early is essential. This is why any vegetable grower worth their salt must have a space dedicated to seedlings. Home gardeners, on the other hand, can not always justify investing in a greenhouse, due to the small number of seedlings they would produce. Instead, they can convert a space in their home to a seedling room: a veranda, an entrance with a glass roof, or even a corner in the kitchen. For home gardeners, and even for microfarms in the start-up phase, this can be a good compromise to minimize expenses. The seedling room must be kept at the right temperature between 64°F and 73°F (18°C and 23°C) during the day and 64°F (18°C) at night and good humidity (60–90% humidity), and above all, it must get a lot of light—14 to 16 hours a day—hence the importance of using supplemental lighting if needed.

Microfarmers growing on more than 2.5 acres (1 ha) will find that a seedling room is insufficient, and having a space dedicated to seeding and growing young plants is a must: In such cases, these are called nurseries. They are heated greenhouses or tunnels that, unlike seedling rooms, are specifically dedicated to seedling production. If you have a greenhouse, the best solution to avoid building a structure that will only be used a few months a year is to devote one part of it to seedlings. All you need to do is lay a geotextile fabric on the ground to suppress weed growth, and set up removable tables. When you are able to seed outdoors and begin planting seedlings into beds, remove this equipment (geotextile and tables) so that you can grow vegetable crops right in the ground.

Dial Seed Sower

With dial seed sowers, also called hand seed sowers and mini hand seeders, seeding is fast and easy. They are especially popular with home gardeners. However, market gardeners also use them to sow small quantities of vegetables.

Purpose and Advantages

Whichever container type you may use for seeding—plug flats, pots, or open flats—the dial seed sower is an essential tool. It allows users to seed with precision. At this stage, it's easy to make the mistake of sowing too many seeds. The result: once they germinate, you end up with too many seedlings which then need to be thrown away. To remedy this problem, the dial seed sower drops only one seed at a time. It consists of a receptacle or cup, which you fill with seeds, a lid with different sized openings, and a small spout that delivers the seeds. You can adjust the lid to select the outlet that matches the seed size.

Dial seed sowers are easy to find at a low price. There's really no reason not to get one!

How to Use It

1 **Pour seeds into the receptacle of the seeder. Close the lid and set it to the opening that matches your seed size.**

2 **Sow the seeds one by one into seed flats, pots, or plug flats that you previously filled with potting mix.**

3 **Cover with a thin layer of soil and tamp down lightly with your hand, then water with a gentle spray.**

Caring for Seedlings

When it comes to seedlings, watering and potting up are two risky operations. They require mastery, patience, and precision.

→ WATER uniformly and according to the size of each seed flat, pot, and plug flat. Make sure that each one gets the same amount of water so that no cell, pot, or tray will dry faster than the others before the next watering. The easiest method is therefore to group your trays by size beforehand. The amount of water you provide will also depend on the location of your cells, pots, or trays within the greenhouse. Those exposed to significant sunlight or a bright light will, logically, need more water. Lastly, temperature is an important factor, both outdoors and in the greenhouse. On warm, sunny days, watering should be more generous and more frequent. Conversely, when the weather is overcast, it is better to space out waterings, or even postpone them.

Watering is always done in two stages: First, moisten the soil so that gravity will allow the water to penetrate the soil surface; with a second pass, you can then wet the entire soil block. Poor irrigation management can lead to disease, especially if the soil is saturated or if the foliage is not drying out properly. Last recommendation: for watering, use tepid water. When it is too cold, it can shock the plants, which will hinder seedling development.

→ POTTING UP is the process of transferring seedlings to larger cells or containers (pots, buckets). When given more space, the roots, as well as the leaves and stems, will grow more as they have a larger and richer soil block.

Potting up is both easy and delicate. You must be gentle when handling seedlings. To extract them, the best method is to pinch the bottom of the cell and grasp the seedling, gently pulling the entire soil block out. If you pot up seedlings at the right time, the soil block will be nice and compact thanks to a network of roots holding the soil together. If the seedlings are weak or diseased, don't bother potting them up. It's better to compost them.

Tools for Seeding in the Field

Unlike the dial seed sower, which is used to sow seeds one at a time into trays, many outdoor seeder models can sow multiple seeds at a time, in one or more rows, and do so much more efficiently. They make it possible to seed faster and with more precision. They use seed distribution technology to avoid seeding crops too densely, which would later require thinning, a tedious operation that involves going back to the rows to pull out excess plants—it can take a long time! With a good vegetable seeder, in comparison, you can quickly achieve the seeding density you need.

Jang Seeder

Designed in South Korea, the Jang seeder is very popular and widely used by vegetable microfarmers. While it is a bit more expensive than other seeders, the cost is offset by its many advantages.

Purpose and Advantages

Easy to use, the Jang seeder requires no special skills. If you've prepared your bed surface properly, adjusted the seeder settings correctly, and use it at the right time, seeding with the Jang is child's play. Adjustable handles make the seeder particularly ergonomic and allow you to walk in the aisle while using it.

The Jang seeder is a high-precision tool. It can be adjusted in many different ways, and features rollers of various sizes to match your seed size and spacing. The rollers are the components that drop the seeds onto the bed.

The seeder primarily consists of transparent hoppers, which are the containers that hold the seeds, an adjustable brush to control seed output, and a drive wheel. As it turns, the wheel drives the rotation of the rollers. The seeds are then fed into the rollers and directed towards a furrow, which is dug by (and also filled by) the shoe. Lastly, the press wheel at the back of the seeder firms the soil for better contact between seed and soil, which guarantees good germination rates.

The single-row seeder, ideal for home gardeners, is the least expensive. For larger areas, 2.5 acres (1 ha) and up, it is better to opt for the Jang seeder featuring 2 or 3, even up to 5, rollers. With this tool, there is no faster way to plant multiple rows at the same time! Since the hoppers are quite large, they do not need to be reloaded often.

How to Use It

1. Prepare the bed and clear away any debris. Pour your seeds into the hoppers, then adjust the rollers for your seed size.

2. Push the seeder down the bed. For consistent seed density, make sure to maintain a constant speed.

Tip from Jean-Martin Fortier

Jang seeders are less delicate than other types of seeders. With good bed preparation, they also provide a considerable advantage: they can be used on a range of soil types. Before seeding, it's best to work the soil with a tilther, hoe, or power harrow and to use the bed preparation rake to rid the surface of debris and stones that could get stuck in the seeder. The beds should also be quite dry.

To be sure, grab some soil, press it into a ball, and drop it from shoulder height: If it breaks when it hits the ground, it's dry enough for seeding.

Six-Row Seeder

The Six-Row Seeder was developed for the type of dense, tightly packed seedbeds seen on microfarms. It is ideal for sowing carrots, spinach, mesclun, and radishes.

Purpose and Advantages

Developed by Eliot Coleman (see p. 2), the Six-Row Seeder is suitable for small and medium-sized seeds. It features six hoppers, set roughly 2.5 inches (6 cm) apart, which makes it possible to sow rows so close together that a hoe will not fit in between the rows. Luckily, this density limits weed germination.

The seeder consists of a front roller that levels the soil, a rotating shaft that dispenses the seeds, and a rear roller that closes up the furrow. Seeding depth is easy to adjust, so it can sow all types of seeds, from small to large. While this seeder was designed for densely spaced crops, it will sow with wider spacing when seeds fill every other hopper. Lastly, you can even combine several different vegetable crops on the same bed by filling each hopper with the seeds of your choosing.

How to Use It

1 **Remove all debris from the bed. Level and compact the soil surface.**

Pour your seeds into the hoppers and adjust the handles so you can walk in the aisle while seeding.

2 **Run the seeder over the bed at a consistent speed, then water the bed with a gentle spray.**

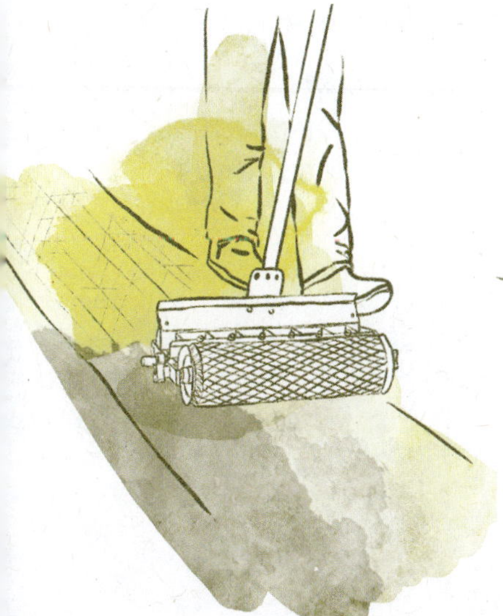

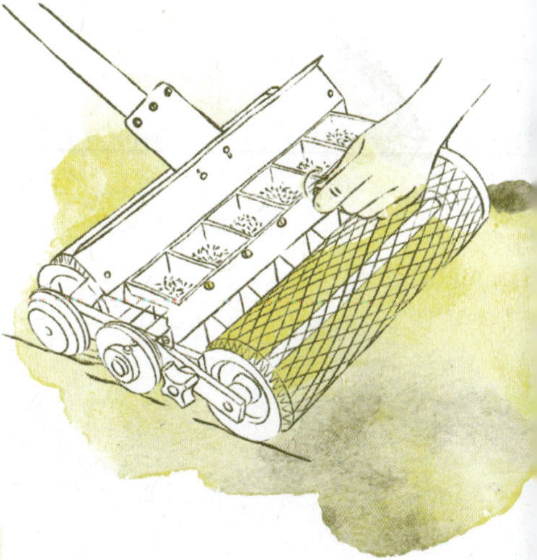

Seedbed Roller

Although it appears simple, this tool breaks up clods, smoothing and firming the soil surface to create a seedbed that supports good germination rates.

Purpose and Advantages

The success of a crop depends largely on bed preparation. Running a seedbed roller down the bed is the ultimate finishing step. It levels and smooths the soil by breaking up the last clumps of dirt. You can do without this tool if you use a power harrow that also tamps down the soil.

The seedbed roller can also be used as a row marker by adding dibbers to the roller, with adjustable spacings. These spikes mark the soil, indicating where to plant your seedlings.

How to Use It

1. Position the seedbed roller on a bed that has already been loosened and smoothed out.

2. While walking in the aisle, push the roller down the bed in a straight line. Go over the bed a second time, if needed.

Tip from Jean-Martin Fortier

The seedbed roller provides the best possible seedbed. By firming and breaking up the soil, the roller optimizes contact between soil and seed. It promotes good germination for even the smallest vegetable seeds.

Planting Equipment

Planting, also called transplanting, marks the end of the first growing stage (growing seedlings in the nursery) and the beginning of the second stage, which starts in the field. Vegetable seedlings, grown in flats or pots, are hardened off outside and later planted in permanent beds. Before planting, the soil is loosened and leveled, and then, depending on the type of vegetable and the spacing it requires, rows are marked on the planting surface. Using a row marker avoids the need to stake a string down the length of the bed, which is tedious. Most market gardeners use a traditional transplanter to make the hole and pour the soil around seedlings pulled from a plug flat. But a small revolution is underway with the use of the Paperpot system (see p. 47): Planting is done with a transplanter operated by a single person—highly efficient.

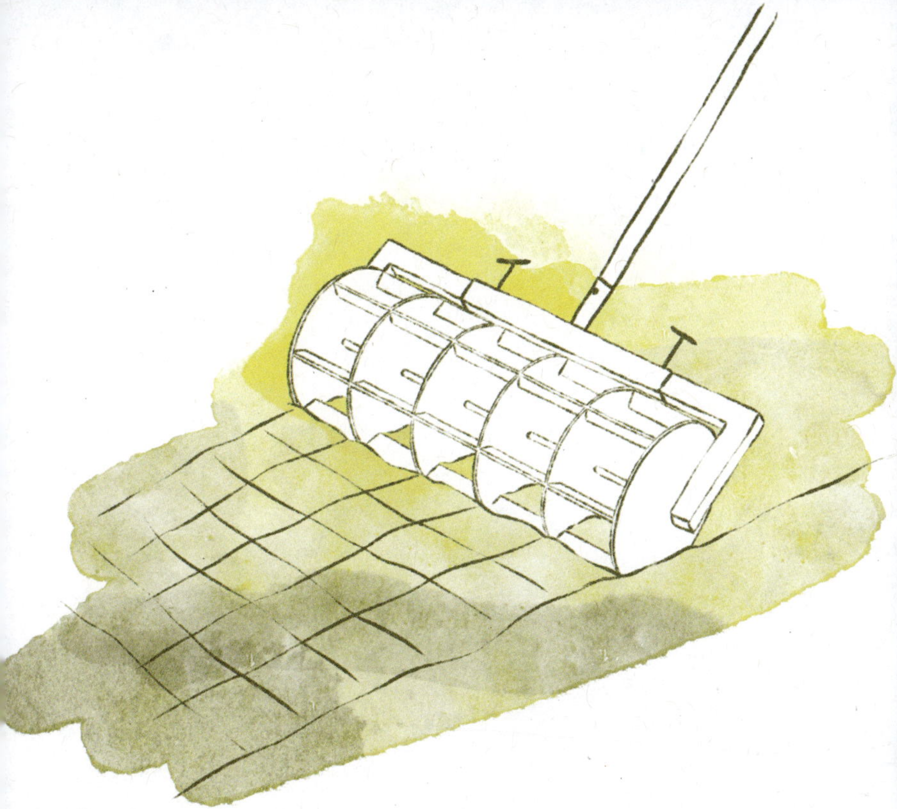

Row Markers

Row markers are highly practical tools that are now replacing the more common method of staking a string down a bed and making it taut. These tools allow growers to work faster and more efficiently.

How to Use It

Row markers are meant to be used in the same way as the seedbed roller (see p. 60).

Purpose and Advantages

Several types of commercial row markers are available. The simplest one is the row marker roller.

→ Row Marker Roller

This roller consists of steel discs (up to 6) connected by 4 horizontal bars. With this tool, growers can draw a grid on the bed surface. The spacing between rows varies depending on the number of wheels, from 14 inches (37.5 cm) with 2 wheels to 5 inches (12.5 cm) with 6 wheels. The roller you choose will depend on the vegetable crop to be seeded or planted.

The handle is purchased separately.

→ Seedbed Roller with Dibbers

This tool is actually a seedbed roller (see p. 60) with dibbers attached. V-shaped, they are clipped onto the roller and allow growers to create holes in the ground or punch through a fabric mulch. When working with heavy soil, you can give the roller more weight by adding one or two sandbags onto the frame.

→ Leek Planter

Also called a leek dibber, this tool looks like a broadfork, with its U-shaped handle and 3 tines. It is specifically used to transplant leeks, resulting in longer white shanks in perfectly straight rows.

The planter also features row markers above the tines, which indent the soil behind the tines, to mark the location for the next row of holes. This makes it possible to achieve rows with highly consistent spacing.

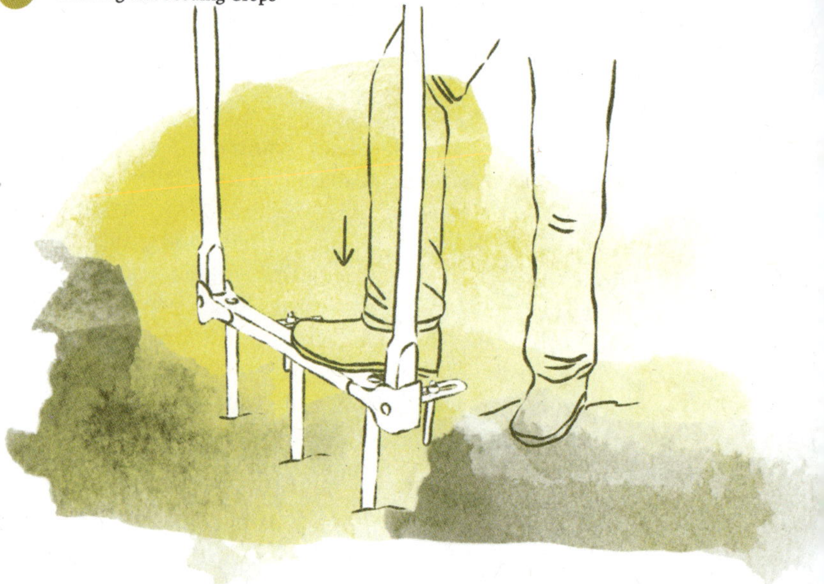

How to Use the Leek Planter

1. Adjust the tine spacing. To make your first holes, drive the tines all the way into the soil, using your body weight and keeping your back straight.

2. Step back to the holes left by the row markers, and repeat.

Tip from Jean-Martin Fortier

The leek planter saves a lot of time, because transplanting leeks is time-consuming. The traditional method is also quite tiring. It requires digging furrows that are 9 inches (24 cm) deep, then filling them back up after planting the leeks, making sure there are no air pockets around the roots. In other words, it's great physical training for athletes!

Garden Trowel

With a short wooden or steel handle and a blade with a slight cylindrical shape, this little shovel is the most commonly used tool for planting seedlings.

Purpose and Advantages

Garden trowels are used to dig holes for transplanting vegetable seedlings.

Models with a narrow, pointed blade are easier to press into the soil and more suitable for seedlings with small root balls; in just a few movements, you can dig a hole, bury the root ball, fill in any gaps, and tamp down the soil. Garden trowels with a wider blade are best suited for planting seedlings grown in pots, as they have a larger root ball, which requires a wider and deeper hole.

Unlike garden trowels, dibbers (also called dibblers) are steel or wooden sticks with a pointed tip. They are designed to make a narrow hole, its depth depending on your choice of dibber. They are used to transplant vegetable seedlings with bare roots, such as onions and leeks.

Tip from Jean-Martin Fortier

Opt for ash handles without knots as this wood is dense and water-resistant. For metal components, choose hardened steel, or even copper, which can contribute to soil health. Pay attention to the shape of the handle, its dimensions, and weight: You must be able to use your garden trowel with dexterity, it should feel natural!

Paperpot Transplanter

This tool, to be used with paper chain pots, has truly revolutionized the process of transplanting seedlings as it speeds up the work and is easy to use. Market gardeners also appreciate how comfortable it is.

Purpose and Advantages

Once the paper chain pot trays are ready to be transplanted (see p. 47), the transplanter will unroll the paper chain pots, dig a furrow, and plant them right into the soil. The tool is ergonomic thanks to adjustable wheels and handles. Although it is a bit expensive, it provides phenomenal time savings. In just 5 minutes, you can transplant 260 seedlings!

How to Use It

1 Once seedlings grown in paper chain pots are ready to be transplanted, prepare the bed and clear it of debris.

2 Position the transplanter at the start of a row that you have already marked with a dibber, row marker, or string staked at both ends of the bed.

3 Use the metal plate provided, and place your paper chain pot seedlings onto the transplanter.

Unwind the end of the chain and place the first few paper chain pots into a furrow dug by the transplanter.

4 Adjust the handles to get a good grip on the tool and keep your back straight.

Stand in front of the transplanter and pull it to plant a row of vegetables.

Tip from Jean-Martin Fortier

If your soil does not allow for the use of this transplanter, you can plant Paperpot chains by hand:

1. Once the bed has been prepared, use a hoe to dig a furrow.
2. Unwind the Paperpot chain and place it in the furrow.
3. Bury the Paperpot with a rake.

Irrigation Equipment

No water means no vegetables! Intensive vegetable production is highly dependent on irrigation, and therefore it depends on your water supply and irrigation systems. First and foremost, you must have a large enough water supply to avoid relying on water provided by the local community. This source of water can be a rain-fed pond, a well, or a spring. Next, the farm must be equipped with a network that can distribute water to your washing stations and—especially—to your beds, both outside and under shelter. Irrigation will be carried out using various systems, through sprinklers and drip tape, depending on the crop. Irrigation systems are a substantial but essential investment. Growers should also consider reusing water, especially water used for washing vegetables.

Factors to Consider When Optimizing Irrigation

WATER QUALITY
For market gardeners, the ideal solution is to have your own pond on the farm. Not only does it help preserve biodiversity, which is highly beneficial for crops, but it also provides a year-round source of water. If you opt to use the city's water supply for irrigation, you will be dependent on the community, regulations, and potential restrictions in the event of a drought.

WATER QUANTITY
You have to estimate the amount of water needed throughout the year, taking into account the number of irrigation lines to be installed, which depends on how many beds you have and how much water they will require.

LOCATION
To set up an irrigation system, you must be very familiar with your land: Is there a natural basin that would be suitable for a pond? Is this location near your beds or far away? Will you have to set up an extensive irrigation system?

COSTS
Installing an irrigation system is expensive. However, this investment quickly pays for itself because it helps to ensure consistent growth in vegetable crops, prevents water stress, promotes good crop health, and helps deliver high-quality harvests.

EQUIPMENT AND MATERIALS
- A pump, gas or electric, and filter to prevent debris from clogging up irrigation lines.
- Pipes that carry water to your blocks of beds.
- Flexible lines that irrigate each bed and vegetable.

Drip Irrigation

With drip irrigation, growers can set up a precise and efficient watering system by distributing water right at the base of each plant.

Purpose and Advantages

Drip systems distribute water slowly and regularly at the base of each vegetable. With this approach, you avoid wetting the foliage, and beds are watered evenly. However, these systems take longer to install, and require close attention to maintain them. You also need to plan ahead to ensure there will be enough stored water to supply the network.

To properly distribute water over a 100-foot (30 m) bed, the best approach is to plan for 4 lines, with holes spaced every 8 inches (20 cm). After getting a crop in the ground, you can send water out through only 2 of the 4 drip lines for the first 2 to 3 weeks. To be safe, it is best to set up emergency lines, available in the event of an equipment failure, and to adapt the flow according to your soil type, especially if it is particularly sandy and permeable. Drip irrigation is not recommended for direct seedings, at least not during the germination phase: in these cases, traditional watering is the best approach.

One of the drawbacks of this system is that it makes it harder to cultivate the soil, for instance when hoeing. If you do not want to move drip lines, you need to proceed with care.

Sprinkler Irrigation

This irrigation system uses sprinklers installed on rods and can cover several beds; it is easy to move as vegetable crops go through their stages of growth.

Purpose and Advantages

There are two types of sprinklers in sprinkler irrigation systems: standard-sized sprinklers and mini-sprinklers. The former is the most commonly used. For 100-foot (30 m) beds, you will require 4 sprinkler heads, which will cover 3 to 4 beds. The major advantage with this irrigation system is that it is quick and simple to install. Plus, it will not need to be dismantled when the beds have to be cultivated.

Mini-sprinklers are mainly intended for low tunnels and caterpillar tunnels. You'll need to install roughly 20 of them to cover a 100-foot (30 m) bed. They share the same drawbacks as standard sprinklers: They wet crop foliage, which is to be avoided with certain vegetables, especially those in tunnels as such closed spaces promote the spread of diseases.

Tip from Jean-Martin Fortier

To make sure that crops are properly watered, market gardeners can use several instruments to measure and monitor moisture.

The most useful one is the moisture meter, which can be used to determine the soil moisture content. Without a moisture meter, you can assess moisture content by touching the soil or sticking your hand in the ground.

In both cases, you have to use observation and common sense to adjust the amount of water distributed to the crops. Automatic timers and controllers can be added to irrigation lines to program when valves will open and close. Do not forget to turn them off in rainy weather. For market gardening operations, the simplest approach is for one person to be in charge of irrigation, making one or two rounds per day to monitor crop watering.

Crop Protection Tools

Greenhouses, tunnels, row covers, and insect netting all protect crops from inclement weather and provide preventive protection against pest damage. More efficient than the frames used by market gardeners until the twentieth century, greenhouses and tunnels are useful for extending the end of season and getting crops started a few weeks sooner by protecting them from the cold. These kinds of structures also make market gardening easier by providing frames on which growers can hang vegetables in need of trellising, like tomatoes, cucumbers, and pole beans. Row covers, also called floating row covers, can be laid out right on the ground or draped over metal hoops to protect young plants from late frosts. They work similarly to the cloches used by market gardeners in the nineteenth century. Finally, a more recent innovation, insect netting is one of many resources used to naturally protect crops from insect pests.

Low Tunnels

Low tunnels consist of hoops and a plastic film held down by cord. They protect crops from the cold, especially early vegetables.

Purpose and Advantages

The hoops are made of metal, and the cover is a transparent polyethylene film. These low tunnels can withstand winds and some snow loading in winter.

How to Set It Up

1. **Drive the metal hoops into the ground every 3 to 6 feet (1–2 m), making each one span the width of the bed.**

   For perfectly straight tunnels, set up temporary stakes centered at each end of the bed and connect them with a taut string.

2

Lay the plastic film over the hoops.

Run cords over the tunnel, attaching them to the bottom of each hoop, which generally features an eyelet for this purpose. They will press the film against the hoops to strengthen the structure.

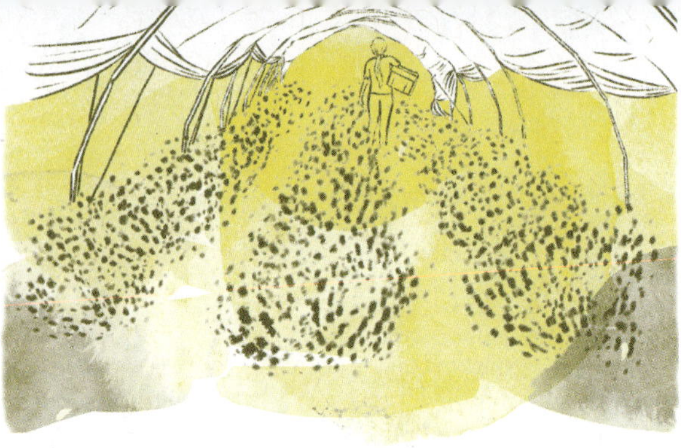

Caterpillar Tunnels

Quick to set up and take down, caterpillar tunnels provide shelter with a decent height; they are just as effective for extending the growing season as they are for sheltering market gardeners!

Purpose and Advantages

Tunnels can be set up at the start of spring to protect early vegetables. Afterwards, they can be used to grow tomatoes, peppers, and melons, which are more sensitive than other crops. In the fall, these same tunnels can protect crops from the first frosts. They provide shelter from bad weather for all—vegetables and market gardeners alike! In hot weather, the plastic film can easily be lifted up along the sides to get airflow around the crops. If a lot of snow is expected, however, it is safer to dismantle them.

How to Set It Up

Caterpillar tunnels are built like the low tunnels described in the previous section. Arches must be spaced roughly 5 to 6.5 feet (1.5–2 m) apart. To reinforce the tunnel, you can tie it down and anchor it into the ground. Alternatively, you can make the plastic taut by running ropes over the tunnel and attaching them to the base of each arch, using eyelets or stakes driven into the ground.

High Tunnels

High tunnels are permanent structures, similar to greenhouses, that provide a beneficial climate for seedlings and delicate vegetable crops.

Purposes and Advantages

These tunnels are an excellent investment for vegetable farms. Despite the high cost, tunnels quickly become profitable since they allow farmers to continue growing crops, and thus increase yields, into the late fall and winter, when nothing grows outside. With a structure made from steel arches bolted together and a polyethylene film, high tunnels are not particularly mobile but are still very easy to assemble.

Typically 20 to 26 feet (6–8 m) wide and 10 feet (3 m) high, they are smaller than traditional greenhouses and come in different lengths, depending on the number of arches. Two doors, at the front and back, as well as openings along the sides, are enough to provide good airflow around the crops. This is a four-season tunnel that protects vegetables from bad weather. It can be used both for early crops in the spring and late crops in the fall. In summer, it is also the perfect space for heat-loving crops like tomatoes. Lastly, a high tunnel is the ideal investment if you want to grow winter vegetables and have year-round harvests and sales.

How to Set It Up

The most important thing to do before building a high tunnel is to select and prepare the location. The best approach is to level the soil and install agricultural drains along the edges to collect rainwater. This will prevent it from accumulating under the greenhouse and saturating your beds. If the beds stay wet for too long, you will have to delay planting early vegetable crops. The tunnel should not be built near tall trees or buildings that would filter the light and cast too much shade.

GREENHOUSES AND HIGH TUNNELS—WHAT'S THE DIFFERENCE?
Unlike high tunnels, greenhouses are heated shelters. They protect seedlings from cold winds and create a beneficial climate for them. They can be made of either glass or plastic. In the latter model, the greenhouse is covered with two layers of plastic film, which are held apart by a blower fan that constantly pushes air in, thus guaranteeing better insulation.

In general, greenhouses are equipped with climate control devices. There are many greenhouse models, but the most common for market gardening are 30 feet wide and 100 feet long (10 × 30 m).

Row Covers

A recent horticultural innovation, row covers act as a shield against the cold, improving germination rates and the hardening off process for seedlings.

Purpose and Advantages

Row covers are made from non-woven polymer fiber fabrics. They create a barrier around plants, protecting them from wind and some insect pests. They also help increase soil temperature and retain moisture. These extra few degrees in soil temperature can be enough to protect crops from frost. Without this protection, vegetable growth slows down when overnight temperatures drop to 46 to 48°F (8 to 9°C), and can even stop entirely around 41 to 43°F (5 to -6°C).

Row covers come in varying thicknesses. Those weighing 0.50 oz/yd² or 0.55 oz/yd² (17 or 19 g/m²) offer a good compromise between lifespan, frost protection capacity, and light transmission because they allow about 85 percent of the sunlight to get through. If you want to create a more effective thermal barrier, simply double the row cover thickness.

Row covers can last about three years if handled with care. Their major advantage is that they are lightweight, inexpensive, and easy to set up.

How to Cover a Direct Seeding

1 Pull the row cover directly onto the bed.

2 You can later add hoops, as soon as the seeds germinate and seedlings appear, to avoid letting the cover come into contact with the plants, which can cause damage.

How to Cover Seedlings

1 Stick flexible metal hoops into the ground to hold up the cover. They will keep it from touching the plants and create an insulating air pocket to further protect them from the cold.

Start by installing hoops every 5 feet (1.5 m). Finish with two hoops set up like an X at each end of the bed, for added strength.

2 Unroll the row cover in the aisle: Cover the hoops and pull the cover taut to prevent the wind from rushing in. Always handle and pull the cover with both hands to avoid tearing it. Place weighted bags around the entire perimeter of the cover to keep the cold out.

Tip from Jean-Martin Fortier

Once spring comes to an end, replace the row covers with insect netting. Roll up the covers and store them in a shelter, away from rodents and bad weather. If kept outside, they may age prematurely.

Insect Netting

Insect netting is a preventive tool to protect crops from insect pests. Simple and effective, it significantly reduces the need for biopesticides and contributes to operating a 100 percent organic production.

Purpose and Advantages

When used at the right time and installed properly, insect netting reduces crop losses caused by insect damage, as well as the need for biopesticides. One very important precaution to take: The net must form a tight seal around the crop. Even the slightest hole is enough to allow pests to get in and take over a bed. So, when installing insect netting, you must be extra meticulous and carefully check that there are no gaps.

Netting can be used on a very wide range of crops. With brassicas like cauliflower, for example, we recommend installing insect netting as soon as they seedlings are transplanted. It should stay on until they become too big and are lacking space. This will protect them from white grubs, cutworms, and cabbage butterflies. Netting can also protect leeks and carrots from flies, and radishes, cabbage, and turnips from flea beetles, for example.

Nets are quick and easy to install. They are quite durable and can be reused several seasons. You can even install them in the summer because they do not affect crop temperatures, let irrigation and rainwater pass through, and also serve as windbreaks. You can choose from a range of mesh sizes, depending on the insect you are trying to keep out.

Plan to install netting soon after planting or, at the latest, when the first insects are detected.

How to Set Up Netting

Insect nets are installed exactly like row covers (see p. 85). Make sure to lay them over hoops to avoid damaging the vegetables.

Once crops are well established, avoid letting weeds take over as they will be in direct competition with your vegetables. The goal is to minimize weed seed germination by suppressing any weeds before they go to seed to prevent their spread. Diligent manual weeding is therefore essential, even if it is time-consuming and requires rigorous planning to integrate it into your task schedule. In the long run, regular weeding will have a positive impact on permanent beds because it reduces the number of weed seeds stored in the soil. The more you weed, the less you'll need to weed!

Finally, it's time to harvest.

Harvesting is the culmination of all your hard work! Like all other tasks, harvesting requires know-how and planning, but also a few tools, so that you end up with high-quality and fresh vegetables as soon as possible.

Crop Management and Harvest

Weeding Tools

Hoes and other hand-held weeding tools have a dual purpose. They destroy weeds that are still in the seedling or cotyledon stage by uprooting them, thus eliminating any potential for recovery. They also loosen and aerate the surface soil layer, which makes it easier for water and nutrients to penetrate and promotes air exchange between the atmosphere and soil. This significantly improves soil life, creating optimal conditions for crop development.

The weeding tool you choose will depend on the type of vegetable crop, its stage of growth, and the spacing between rows. So, it's better to have several models in your toolbox.

Stirrup Hoe

The stirrup hoe is an essential tool for weeding and loosening the soil. Today's models, with interchangeable blades, are suitable for all types of crops and vegetables.

Tip from Jean-Martin Fortier

The art of hoeing is about good timing. You need to hoe beds at the right time, when the weeds are still in the cotyledon stage (after the first two leaves appear from a germinated seed).

The second secret to successful hoeing is maintaining a sharp blade.

Purpose and Advantages

The stirrup hoe features a long wooden handle and a rectangular steel blade. The blade oscillates as the handle is moved back and forth, in a push-pull effect that disturbs the soil surface. The blades come in a range of widths. The narrowest one, at 3.25 inches (9 cm), is best for permanent beds with 4 or 5 rows; the mid-range blade, at 5 inches (13.5 cm), is suited to crops with 2 or 3 rows; and the widest, at 7 inches (17.5 cm), is recommended for weeding single rows and aisles.

Stirrup hoes allow growers to get as close as possible to the crops. Avoid pushing the blade too deep into the soil as this may turn over the layers, bringing weed seeds to the surface.

Ideally, crops should be hoed when the weather is hot and dry. If the soil is moist and the leaves do not dry out in the sun, the weeds might take root again.

Stirrup hoes are quite useful with both direct seedings, like fennel or carrots, and slower transplanted crops, such as cabbage because they eliminate the roots of both well-established weeds and their young seedlings.

A further advantage of the stirrup hoe is that it breaks up the soil crust formed by repeated watering. The soil surface is then more permeable, allowing better infiltration of water and rainwater, as well as any nutrients they contain.

How to Use It

1. Adjust the angle of the blade so that you can keep your back straight while using the hoe.

2. As you walk backwards down the aisle, holding the handle at chest height, push and pull the hoe between the plants.

Collinear Hoe

The collinear hoe is particularly useful when working between the rows of a mature crop because it allows growers to cultivate the soil as close as possible to the plants without damaging their foliage.

Purpose and Advantages

On this hoe, the fixed steel blade ranges from 3.25 inches (9 cm) to 7 inches (17.5 cm) wide and is at a 70-degree angle to the handle. It is the ideal tool for weeding low-lying crops like cabbage and head lettuce.

It can be used when weeds are more established, once they have their first true leaves, but is still most effective to remove weeds in the cotyledon stage.

How to Use It

1 Hold the handle with your thumbs pointing up, back straight, and the handle at chest height. This is the most ergonomic way to avoid becoming overtired.

2 Scrape the entire surface of the bed, with the blade just a half inch (1 cm) deep. Skim the soil under the foliage and near each plant base, being careful not to cut the roots: if weeds disappear as the hoe runs through the bed, the blade is at the right depth.

Tip from Jean-Martin Fortier

For optimal results using the collinear hoe, weeds should be severed between the root and the stem. This is why you should stay right below the soil surface because if you drive the blade too deep, you risk simply going under the roots you mean to cut!

Wire Hoe

The wire hoe, or interchangeable wire hoe, is not well-known among home gardeners but is very popular with market gardeners for its precise and gentle cultivation that doesn't damage plants.

Purpose and Advantages

The wire hoe is similar to the stirrup hoe, except that the tool at the end of the handle is not sharp. The wire itself forms a triangle with rounded sides.

This tool allows growers to work particularly close to plants and drip tape without damaging the crop or irrigation equipment. Unlike other hoes, it does not cut the roots but pulls them from the soil, which reduces the likelihood of regrowth.

How to Use It

1. **While walking backwards in the aisle, with your back straight and the handle at chest height, scrape the soil surface by pulling the hoe towards you.**

2. **By holding the tool upright, you will get as close as possible to the plants and slip between them without causing damage.**

Tip from Jean-Martin Fortier

The wire hoe is more effective with weeds that are still in the cotyledon and, therefore, have weak roots. You will need to use a stirrup hoe on more established weeds. In addition, the wire hoe is only useful for light soil types.

Flex Tine Weeder

Tine weeders are rakes with thin, flexible metal tines that gently and quickly weed and loosen the bed surface.

Purpose and Advantages

Flex tine weeders consist of a wooden handle and one or two rows of flexible, curved tines. As it rakes the bed, the tool scrapes the soil surface and digs up weeds. The tines act as springs, bouncing off the more developed and sturdy vegetable roots, without damaging them. These tools can even be used on creeping roots, or runners, that have already spread significantly. They are safe to use on crops, allowing growers to rake an entire bed very quickly and without precision. For efficiency, it is best to choose a weeder that is the same width as your permanent beds (30 in, 75 cm). Just like the hoes, tine weeders break up soil crusts caused by repeated watering. They are quite convenient for very dense plantings, on beds with 8 to 12 rows, for example. Above all, tine weeders are preventive tools that have to be used regularly for maximum efficacy.

For direct seedings, where the soil is not protected by crop foliage, it is best to run the flex tine weeder over the bed once a week, until the seedlings have become well established and can compete with weeds.

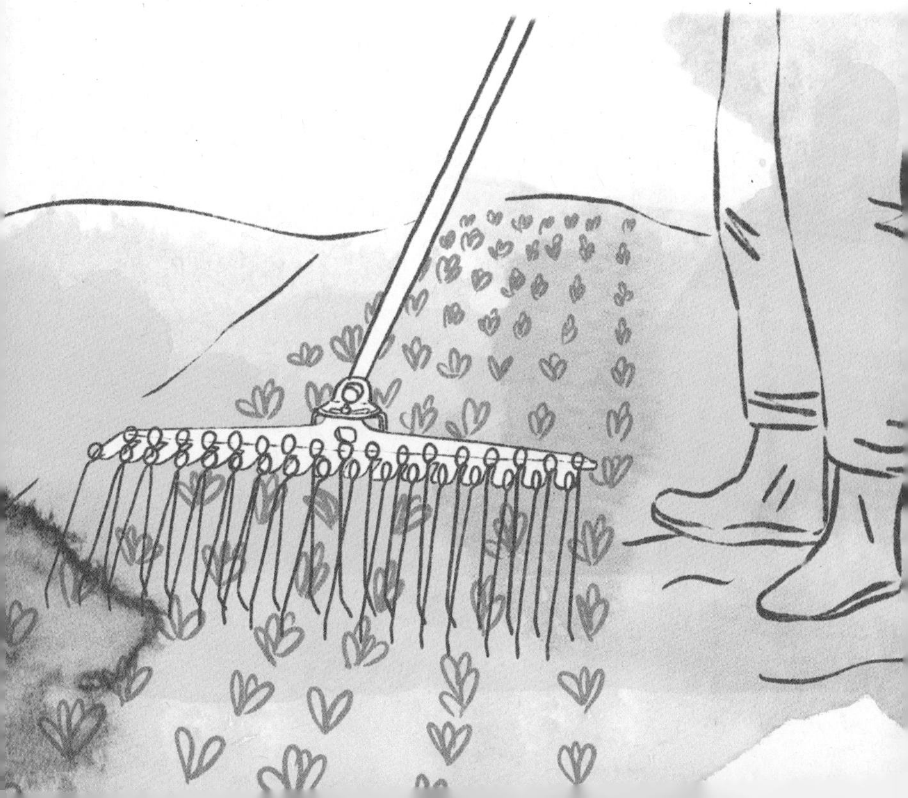

How to Use It

1. Place the tine weeder in the middle of the bed. Hold the handle at chest height and keep your back straight.

2. While walking backwards down the aisle at a consistent pace, pull the rake down the entire bed. Once the crop is well established, you will be able to apply more pressure.

Tip from Jean-Martin Fortier

To determine the right time to use the flex tine weeder, the easiest method is to test it at the end of a bed and see what happens. If the weeder damages your crop, wait a few more days!

The flex tine weeder can be used to weed crops with fairly fine foliage, like carrots, fennel, and onions. The sharper the angle between the weeder and the crop bed, the more aggressively it will weed. For best results, avoid using it on wet ground. This is why we recommend using this tool on a sunny, hot, and dry day.

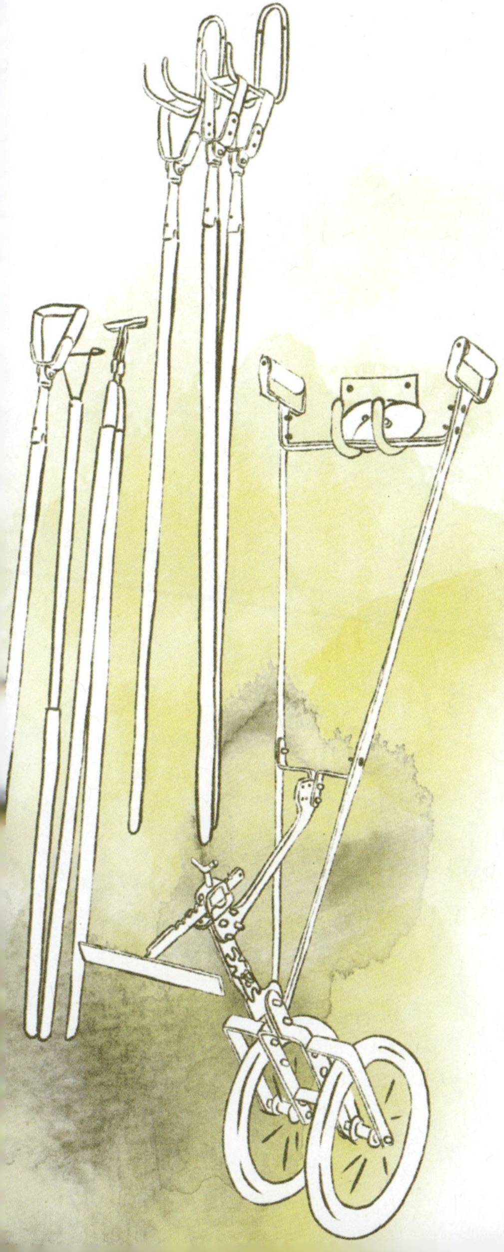

Weeding

Weeds are not harmful by nature. In general, they are in fact quite useful for biodiversity. However, in crop beds, they are in direct competition with the vegetables. They absorb nutrients, water, and light while taking up the space your plants require.

To limit their spread, prevention is your best option. Some highly effective proactive solutions were mentioned above:

- Plant tight rows to prevent the sun from reaching weeds.
- Work the soil surface without turning it over to avoid drawing up deeper weed seeds.
- Eliminate any weed seeds in your amendments.
- Mulch the ground.

Still, despite all these precautions, weeds are pugnacious and manage to get established. This is why beds need to be hoed regularly.

Bio-Discs

Bio-discs are an accessory designed to be attached to a wheel hoe (see p. 24). They allow growers to weed with precision, while loosening the soil and hilling crops.

Purpose and Advantages

Bio-discs consist of two pairs of discs that attach to a wheel hoe; the inner discs are small and straight, while the outer discs are large and parabolic. Their main advantage is you can weed in the row while hilling a crop.

They therefore perform two tasks at once that would otherwise require multiple tools, thus saving time. What's more, unlike the walk-behind tractor, the wheel hoe is completely silent. The wheels and two handles make it very easy to use, with little effort, and it is quite durable. This tool makes it possible to hoe large areas quickly. The only prerequisites for using bio-discs are that rows must be straight and the soil should be fairly light.

How to Use Them

1. **Adjust the bio-discs according to row spacing in the bed to be weeded.**

2. **Place the two wheels of the wheel hoe and the two pairs of bio-discs on either side of the row. Push the hoe in a straight line, without damaging the plants.**

Tip from Jean-Martin Fortier

Designed for user comfort, the wheel hoe does multiple tasks at the same time, is ecofriendly, and is ergonomic. It's a must-have tool for every market gardener!

Other Tools

A few additional tools will complete the market gardener's kit. These may support some market gardening operations but are not necessarily indispensable. The list includes compost spreaders, which are not described in this book, and flame weeders, which are used infrequently and thus not always a justifiable investment. As for hand sprayers and other sprayers with manual or battery-powered pumps, they are more commonly used on microfarms to spray fertilizing solutions made from plant-based manures or to apply organic pest control products—only after preventive actions have failed. We strongly recommend having two separate devices for these uses.

Hand Sprayer

With a capacity of just 1 to 2 quarts (1–2 L), this mini sprayer, whether it has a trigger or pump mechanism, is intended for occasional use by home gardeners.

Purpose and Advantages

The most common sprayer (shown above) has no system to pressurize the contents. Other models have a pump mechanism that pressurizes the liquid in the tank. This allows users to spray continuously, without having to constantly squeeze the pump handle, instead focusing entirely on applying a fertilizer or pesticide solution. The spray nozzle is adjustable and can distribute a fine mist that will cover the entire plant.

For home gardeners, who only occasionally spray plants, this model is sufficient. Professional market gardeners, however, will opt for a backpack sprayer that has a larger tank for greater autonomy.

Backpack Sprayer

Backpack sprayers make it easy to apply biopesticides and plant-based fertilizers over greater surface areas.

Purpose and Advantages

The backpack sprayer consists of a tank that can be carried on your back using shoulder straps, a pump, and an extension wand to spray as close as possible to the plants.

With this pump, users can distribute a range of liquid products, biopesticides, or plant-based and algal solutions. That's why we recommend having at least one of these sprayers in your toolbox. It's best to buy a good-quality model that will last longer, even if it means paying a little more. To get the most out of your sprayer, it's important to use the right dosage and to follow the recommendations for timing, before and after watering, for each application. Make sure to apply sprays on dry and overcast days, not during rain.

The backpack sprayer can be used to apply biopesticides. These natural plant protection products do not rely on chemical processes, are biodegradable, and will not leave chemical residues in the soil.

They may address one or more types of insects, temporarily or over several days. Use these products sparingly, and only in cases of heavy infestations: The aim is to control insect pest populations, not necessarily to eradicate them. When it comes to biopesticides, less is more!

How to Use It

1. **Wear the protective equipment recommended for the product you are applying (goggles, gloves, mask, etc.). Use the pump to pressurize the tank, then spray while walking backwards to avoid any contact with the product.**

2. **While you are applying the product, shake the tank from time to time to keep the diluted content from settling. When you are done, rinse the tank, extension, and nozzle several times with clean water. Then, store the sprayer in a dry and dark space.**

Tip from Jean-Martin Fortier

Some models pressurize the liquid with a battery-powered pump. Although more expensive, this option provides a more pleasant, consistent, and comfortable spraying experience because it requires no manual pumping.

Flame Weeder

A flame weeder kills weeds that have recently germinated or have a few leaves. It is used just before some direct seedings.

Market gardeners have access to many tools that help with weed control. Before using a flame weeder, which is powered by propane, a fossil fuel, we recommend applying a few preventive principles:

- Plant rows very close together.
- Never turn over the soil.
- Apply weed-free compost.
- Use woven mulches.
- Cover beds with silage tarps
- Use the stale seedbed technique before using a flame weeder (see Jean-Martin Fortier's tip below).

Purpose and Advantages

A little training is required before using a flame weeder, but it is fairly simple to use.

The tool consists of torches, a flame cover for wind protection, and a frame supporting the entire apparatus and fuel supply. With wheels guided by handles, it is easy to move around. The flames can therefore have a targeted effect on the surface you are addressing. Weeds are destroyed in seconds.

For best results, the surface must be flat because uneven soil may deflect the flames. Weeds are most effectively controlled when they are still in the cotyledon stage or have only their first true leaves, when they are less than 2 inches (5 cm) tall. The flame weeder is especially useful for vegetables that germinate slowly, like carrots and beets. For weeds that germinate quickly, we recommend running the flame weeder over the bed once before seeding the crop. However, it should be noted that given the cost, this is an investment that can be delayed.

Tip from Jean-Martin Fortier

The best way to use the flame weeder is to first apply the stale seedbed technique. This involves preparing the seedbeds a few weeks in advance so that weed seeds have time to germinate, which makes them easier to eliminate. You can then wipe them out with a power harrow or wheel hoe. The flame weeder comes in at this stage, sometimes called pre-emergence flame weeding, and here's how to go about it:

- **Prepare the bed one week before direct seeding.**
- **Amend and water the seedbed. You can use row covers to encourage weed growth.**
- **Once weeds have germinated, use the flame weeder.**
- **One week later, sow your vegetable crop.**

Harvest Tools

For market gardeners, harvest time is very gratifying because it culminates several months of cultivation and constant, attentive care. Vegetable harvesting techniques, involving a multitude of small repetitive motions, must be streamlined to eliminate any actions that are unnecessary, tiring, and not ergonomic. In addition to rigorously organizing harvesting steps, growers need to use appropriate equipment—this is essential. From hand tools, such as pruning shears and knives, to harvest carts, crates, bags, and more sophisticated motorized tools like greens harvesters—all equipment contributes to improving comfort and streamlining tasks for market gardeners. This, of course, will affect the profitability of your microfarm.

Quick Cut Greens Harvester

This tool was specifically designed to quickly harvest delicate baby greens and, above all, to do so as carefully and gently as possible.

Purpose and Advantages

The Quick Cut Greens Harvester consists of a metal shaft with a sharp horizontal blade and macramé brushes that resemble the spinning brushes used in car washes. The brush pushes the cut leaves back into a canvas basket.

The Quick Cut Greens Harvester is powered by a simple drill that drives the mechanism as it rotates. For best performance, opt for an 18-volt drill.

It can also be used to harvest a wide range of greens, including mesclun, mustard greens, arugula, spinach, and baby kale.

How to Use It

1

Place your harvest bins at one end of the bed.

Hold the harvester with a firm grip, keeping it parallel to the ground.

Turn on the drill to get the brush spinning and drive the blade, which will cut the leaves while leaving the new growth at the heart of the plants intact. For cut-and-come-again crops, this means they will be able to grow back, to be harvested again.

2

With an 18-inch (45 cm) cutting width, the harvester can only cover half a bed, which is typically 30 inches (75 cm) wide. You will therefore need to harvest in two passes, first down one half of the bed, then turning the machine around to cut the second half. To do this, lift the harvester without stopping it, so that the brush continues to feed leaves into the canvas basket. Then, begin cutting the remaining half of the bed.

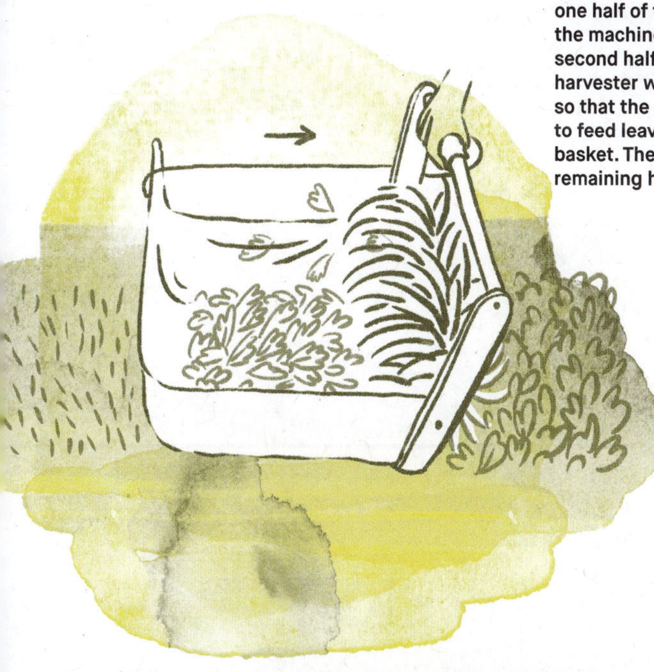

Tip from Jean-Martin Fortier

The Quick Cut Greens Harvester is a brilliant invention. Compared to the traditional method of using a knife, this tool harvests greens ten times faster, with an unbeatable output of 175 pounds (80 kg) per hour. The only prerequisite for good results is that your crop needs to be perfect. Because the tool is lightweight, it's easy on the back. For maximum efficacy, make sure that the blades are sharp, and keep spare parts (drill batteries, belts, blades, etc.) on hand to avoid interrupting your harvest, and reverting to cutting greens with a knife!

Harvest Knife

The harvest knife is a versatile and indispensable tool that every market gardener should have in their pocket. Use it for harvesting and cutting or slicing vegetables.

Purpose and Advantages

For harvests, this knife can just as easily be used to cut roots, leaves, or stems. It is most often used for zucchini, leeks, and greens, especially mesclun, when growers do not have a Quick Cut Greens Harvester.

Opinel knives, made in France, are of high quality. When folded, they are very compact. And, thanks to a patented blade locking system, they can be used safely. They are also highly resistant to corrosion.

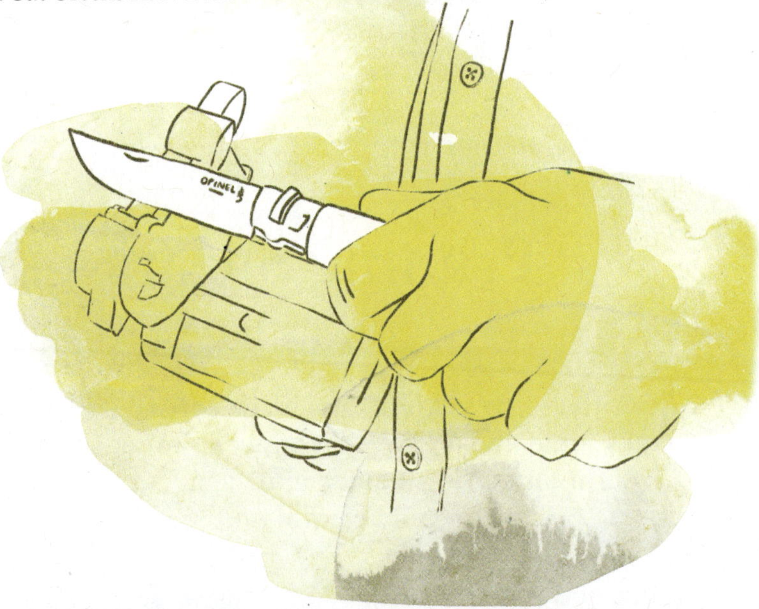

Tip from Jean-Martin Fortier

There is nothing worse than a dull knife, both in terms of the user's safety and the blade's effectiveness. To maintain your harvest knife, the best practice is to sharpen it before every use. This simple action barely takes a minute.

Of all the existing models, I recommend the Opinel No. 10 because it is easy to grip, and the blade is the perfect size for an efficient and clean cut.

Pruning Shears

After the harvest knife, this is the second cutting tool you should always have on hand! Pruning shears are useful for getting a quick and clean cut whenever your knife cannot do the job.

Purpose and Advantages

Pruning shears can make more difficult cuts, complementing work done with a knife. Of the many available models and manufacturers, the Swiss-made Felco 310 is the ideal tool. It is durable and has straight blades that allow users to work on all types of plants, from vegetables and herbs to shrubs and fruit trees. It can also be used to harvest vegetables with tough stems, like peppers, eggplants, and artichokes, and for crop maintenance.

Harvest Bins

Although it may seem trivial, harvest bin selection is important because it will shape the entire vegetable handling and storage system.

Purpose and Advantages

The harvesting process is often quite tiring; it can cause some discomfort and even pain. To avoid this, it's best to select ergonomic and practical accessories.

Sturdy plastic harvest bins can be used for greens, while other vegetables can go into perforated bins. Make sure you choose models that stack perfectly so that you can store them, empty or full, in cold rooms, storage spaces, vehicles, and even at markets.

Harvest Bags

Harvest bags, a recent invention replacing the traditional wicker picking baskets, are both light and flexible, which makes them gentler on both the grower and the harvested vegetables.

Purpose and Advantages

The bags are similar to saddlebags, usually in pairs to balance the load, and they make harvesting much easier. They are connected to straps that distribute the harvest load evenly across the shoulders, without straining the back. Users can work while crouching, kneeling, or standing, keeping their hands free. They don't have to bend over to unload the harvest: The saddlebags open from the bottom, thanks to two clasps. This makes it easy to transfer the contents of the bag, which optimizes harvesting time.

Harvest Cart

Ecological and essential, this equipment makes it easier to transport crops (as well as various supplies) between areas of the farm. It should be one of your first investments!

Purpose and Advantages

Unlike standard wheelbarrows, harvest carts can handle very heavy and bulky loads. But you don't have to hit the gym to be able to pull it! Thanks to its two wheels, anyone can move it with minimal effort. And with them set 30 inches (75 cm) apart, the cart can be pulled down the length of a bed without compacting the soil or crushing the crop. With two metal feet at the front, it is stable when stationary.

Tip from Jean-Martin Fortier

A harvest cart can transport heavy loads, like loose vegetables, full harvest bins, compost or other loose material, and even tools. It can also be used as a mobile workbench, to process or package vegetables right by the beds.

Acknowledgments from Jean-Martin Fortier

I wish to thank the entire team at the Market Gardener Institute for encouraging me to pursue my mission, every day. A big thank-you also goes out to the Growers & Co. team, who pushes me to come up with new types of equipment! I especially want to acknowledge my partner, Maude-Hélène Desroches, who is an exceptional market gardener and a dear friend!

Acknowledgments from New Society Publishers

We extend a great thanks to Delachaux et Niestlé, the French publisher, for working with us to publish this English edition. Further thanks to the New Society Publishers team for producing the book and especially to Laurie Bennett for her meticulous attention to technical details and high-quality translation into English.

Acknowledgments from Delachaux et Niestlé

A big thank-you to Jean-Martin Fortier and his team at the Market Gardener Institute for this wonderful collaboration.

Our heartfelt thanks go out to Pierre Nessmann for putting us in touch with Jean-Martin, for thoroughly editing this collection, and for being so generous with his time. For this book, we owe him so much. We also wish to thank Flore Avram, whose illustrations give this collection a beautiful, simple character; to Grégory Bricout for graphic design that cleverly reflects Jean-Martin Fortier's spirit; and to Sandrine Harbonnier and Sabine Kuentz for their work on the text.

Reference Books

The Market Gardener: A Successful Grower's Handbook for Small-Scale Organic Farming, New Society Publishers, 2014.

Microfarms: Organic Market Gardening on a Human Scale, New Society Publishers, 2024.

Grower's Guides from the Market Gardener

Tomatoes: A Grower's Guide
Vegetable Garden Tools: A Grower's Guide

Coming Soon

Root Vegetables: A Grower's Guide
Living Soil: A Grower's Guide
Fruit Vegetables: A Grower's Guide
A Year of Vegetables: A Grower's Guide

Translator : **Laurie Bennett**

About New Society Publishers

New Society Publishers is an activist, solutions-oriented publisher focused on publishing books to build a more just and sustainable future. Our books offer tips, tools, and insights from leading experts in a wide range of areas.

We're proud to hold to the highest environmental and social standards of any publisher in North America. When you buy New Society books, you are part of the solution!

At New Society Publishers, we care deeply about *what* we publish — but also about *how* we do business.

- This book is printed on **100% post-consumer recycled paper**, processed chlorine-free, with low-VOC vegetable-based inks (since 2002)
- Our corporate structure is an innovative employee shareholder agreement, so we're one-third employee-owned (since 2015)
- We've created a Statement of Ethics (2021). The intent of this Statement is to act as a framework to guide our actions and facilitate feedback for continuous improvement of our work
- We're carbon-neutral (since 2006)
- We're certified as a B Corporation (since 2016)
- We're Signatories to the UN's Sustainable Development Goals (SDG) Publishers Compact (2020–2030, the Decade of Action)

To download our full catalog, sign up for our quarterly newsletter, and to learn more about New Society Publishers, please visit newsociety.com.

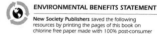

ENVIRONMENTAL BENEFITS STATEMENT

New Society Publishers saved the following resources by printing the pages of this book on chlorine free paper made with 100% post-consumer waste.

TREES	WATER	ENERGY	SOLID WASTE	GREENHOUSE GASES
14 FULLY GROWN	1,100 GALLONS	6 MILLION BTUs	49 POUNDS	6,300 POUNDS

Environmental impact estimates were made using the Environmental Paper Network Paper Calculator 4.0. For more information visit www.papercalculator.org

Certified B Corporation

SDG PUBLISHERS COMPACT

FSC www.fsc.org MIX Paper | Supporting responsible forestry FSC® C016245

new society PUBLISHERS www.newsociety.com